# *Goodbye* Green

# *Goodbye Green*

## How Extremists Stole the Environmental Movement from Moderate America and Killed It.

Glen Duncan

**WBTR-TV
On Assignment, Inc.**
Baton Rouge, Louisiana

**MERRIL PRESS**
*Bellevue, Washington*

# *Goodbye* Green

This book is copyright © 2000 by Glen Duncan. All rights reserved. No part of this book may be reproduced in any form or by any electronic or mechanical means including information storage and retrieval systems without written permission, except in the case of brief quotations embodied in critical articles and reviews.

First Edition
Published by Merril Press
April, 2000

Merril Press
P.O. Box 1682
Bellevue, Washington 98005
Telephone: 425-454-7009
Visit us at our website for additional copies of this title ($14.95 each) and others at www.merrilpress.com

Library of Congress Cataloging-in-Publication Data

Duncan, Glen, 1958-
 Goodbye green: how extremists stole the environmental movement from moderate America and killed it/Glen Duncan.--1st ed.
   p. cm.
 Includes bibliographical references and index.
 ISBN 0-936783-26-5 (pbk.)
  1. Green movement--United States. 2. Environmentalism--United States. I. Title.

GE195 .D85 2000
363.7'054'0973--dc21

00-037975

Printed in the United States of America

# DEDICATION

Goodbye Green is dedicated to all those enviro moms, dads and kids and grassroots grannies and granddads who worked so hard to make our earth a better place, but have retreated to more satisfying and fruitful activities. I'm sorry to see you go, but I understand why you left the movement.

# Contents

Introduction — 13

Quick Note on Use of Examples — 15

Chapter 1. A Simple Little Movement — 17

Chapter 2. The Death of the Movement — 31

Chapter 3. The Price of Success — 43

Chapter 4. Labor goes Green — 53

Chapter 5. Don't mention birds! — 59

Chapter 6. Green Justice — 79

Chapter 7. Global Green — 103

Chapter 8. Other Things That Hurt the Movement — 121

    Data Slinging
    Spreading the Blame
    Superfund
    Agents Provocateur

Chapter 9. Wrap Up — 157

Sources — 161

Index — 169

About the Cover — 179

# Acknowledgements

No writer can do it without a copy editor/supporter and Dre' Tilley (an emerging writer herself) is that. Thank you Dre'. The enthusiasm of my parents, Grey and Glenda Duncan, for anything I've attempted has been unflagging and includes this book. Thanks, of course, to my publisher for seeing something in the manuscript worthy of his time and money. And to my wife, Karen, and children, Taylor and Hillary, thank you for the time you let me sneak off to write.

# Introduction

I covered the environmental movement as a reporter; cut my journalistic teeth on it in fact, and now I miss it. I miss it, and I want it back. Not because I am in agreement with its principles or even in agreement with many of its members. It's just that to me, the grassroots environmental movement represented democracy – revolutionary democracy – warts and all.

Fresh out of graduate school in 1982, I began professional life as a geologist for Amoco Production Company in New Orleans but soon discovered I wasn't that good at finding oil. However, I enjoyed writing the reports about all the oil my friends were finding. In 1986, I hung up my hard hat and grabbed a typewriter to become a general assignment and science reporter for some local newspapers while working toward a degree in journalism. Then I discovered TV reporting. I was lucky to land a job at the ABC affiliate in Baton Rouge, but was the odd bird in the news room. I had spent much more time in the classroom than on the streets, but my boss wanted someone with a science background to put on the environmental beat So I became our station's "environmental specialist" at a great time. Moderates were swelling the ranks of the grassroots environmental movement, making significant strides and grabbing lots of political attention at every level.

While covering the movement, I came to the conclusion that most of this nation's pollution problems could have been solved quickly and easily if calm, reasonable environmentalists would have sat down with calm and reasonable industrialists. Unfortunately, these two foes could reach no truce in their battle for green space and turned to government to mediate. When government did step in, it spawned one of this nation's greatest boondoggles for litigation attorneys. It was known as Superfund, and it tried to pin the clean up of hazardous waste sites on the unsuspecting, including small businesses and even a church retirement home. Let me tell you, it wasn't fun watching the feds use Superfund to punish an old folks home when industrial managers, government officials and activists could have sat down to figure this stuff out reasonably. But just when it seemed environmentalists and industrialists might be learning to talk to each other, the environmental movement died. It was as though the whole movement stepped off a cliff. Where did it go?

Read on. You'll learn about the break up of the movement caused by a fight for membership and money by trade unions, civil rights groups, conservation groups and environmental extremists who tried to capitalize on the groundswell of support from middle America. You'll even get a little insight to how the moderate agenda to make us all responsible for pollution clean-up actually hurt the movement, and how corporate America resorted to some intrigue of its own to shirk responsibility. Perhaps this short book will cause enough discussion to resuscitate the movement, but in a "kinder, gentler" manner.

I can assure you that many people, including my environmental friends, will call me nuts. Many will shoot down my ideas. They will call me blind or uninformed. They'll accuse me of harming the very movement I just said I wanted back. And they'll line up to prove me wrong. To all of these verbal darts, I am ready with a single question: Where is the movement – the real grass roots movement filled with fathers, housewives, grandmothers and students? If you can show me, I'll take back what I am about to say. But I'm not counting on it. This is my story and I'm sticking to it.

# Quick Note on Use of Examples

Before I begin, let me offer an explanation for my use of examples from Louisiana. First, my beat as an environmental reporter was in Louisiana. I'm familiar with the area and have closer personal knowledge of the characters involved. But more than that, Louisiana industries manufacture upwards of 25 percent of our nation's chemicals, making the miles and miles of chemical plants stretching from New Orleans to Baton Rouge one huge target of the environmental movement. Louisiana residents along the Mississippi River didn't just represent a microcosm of the grass roots movement, they were among its leaders, attracting attention and admiration nationwide.

That said, let me be quick to follow with this thought: our country has plenty of excellent, motivated and informed grass roots environmentalists whose stories could be told here. Although this book will stop off at Niagara Falls, Philadelphia, Love Canal, Rochester, Athens, Georgia and other locales and discuss environmentalists from across the nation, please forgive me for not including all our nation's figures and allowing me to use Louisiana personalities to illustrate my point. Thank you.

# A Simple Little Movement Is Born and Goes to Washington

Many journalists, activists and even some historians mark April 22, 1970 as the beginning of a simple, extraordinarily effective grassroots movement in this country. On that spring day, scores of enlightened and motivated citizens all around the country drove, walked or bicycled to their nearby beaches, lakes, parks and ditches to pick out old tires, trash and litter in a great, continent-spanning, community effort. It was the first Earth Day, organized mainly by college students in response to the growing concern over air and water pollution. Slogging through the knee-deep mud of clogged urban streams amid the soggy piles of their reclaimed trash, those citizens had a vision of the entire population getting involved, of politicians working publicly to clean up the environment and industry executives willingly following the lead to clean up America. They darn near got what they were hoping for. During the next decade, environmentalism became a true household word complete with home recycling "centers" in our kitchens, student save-the-earth projects cluttering our dining room tables and scores of programs and news stories on TV and in newspapers charting the progress of our collective clean up effort. Environmentalism actually began much earlier, but had several false starts before building the momentum needed to move a great mass of Americans to action.

The conservationists of the 1890s and early 1900s made us aware of our country's vast wildernesses and beauty by fighting to create national parks. But conservationists could not really stop the development and westward growth that had been called our "manifest destiny." Much of the virgin land was plowed under or built up, and conservationists realized that they had pretty much saved all the unspoiled territory they could. Yet, a yearning for the good earth remained. By the dawn of Earth Day, 1970, a new strategy for pure, clean land began to emerge, pushed by a new generation of earth watchers. These were the environmentalists. Their idea was this: since we as a nation weren't able to save all of our virgin land from the plow, the bulldozer and the industrial developer, then we should go back and undevelop some; clean it up, restore it and let mother earth resume ownership of it. Thus our collective consciousness swung from the idea of stopping development to the idea of reversing it - from preserving to reclaiming.

In the first days of this new environmentalism, we as a society weren't sophisticated enough to know about or worry about the specifics of unseen groundwater contamination, injection wells, incineration, landfarming or landfilling. We simply wanted clean air, land and water, and the event which may have been the first grassroots uprising was sparked by a very basic want for the first of these - clean air. In 1948, residents of Donora, Pennsylvania were forced to make their way through thick, poisoning smog for days. U.S. Steel mills were continuously belching nauseating smoke which soon settled on the town, making residents sick. The residents responded with an outcry and a flurry of lawsuits. U.S. Steel eventually shut down and left town. Nearby Pittsburgh took notice and rewrote its air standards in perhaps the first step in a movement which would ultimately result in the Clean Air Act years later.

Yet the environmental movement had a hesitant start. Opinion polls and citizen actions raised the priority of the environment several times only to have it bumped down each time; first by unemployment, then the Vietnam War, the peace movement, the oil embargo of the 70s, and Watergate's revelations about the thinning integrity of government. In fact, 50 years passed before anyone even remembered the Donora smog, recognized the subsequent civil action as a milestone in the movement and raised the incident to national recognition. Justin Shalley, a student at Ringgold High School in nearby Monongahela, Pennsylvania, held a memorial service for the event in October, 1998 and applied for an historical marker to permanently footnote it. And just about the time of the Donora smog, Aldo Leopold wrote his book *A Sand County Almanac*, becoming one of our first environmentalists. But the next real rallying cry was not until 1962 when Rachel Carson warned in her book *Silent Spring* about the abundance of pesticides seeping into the roots, diets and bodies of plants and animals. Seven years passed before the next stab of our public conscience by the 1969 Santa Barbara Oil Spill which also raised a cry of "foul" from those concerned about clean water and beaches. Earth Day followed that in 1970. Perhaps because it was tied so closely to a very strong, national peace movement, that first Earth Day gained instant and prominent play in the media and in our nation's capital. Unlike the events before, this one may have really gotten things going. The Environmental Protection Agency was formed in October of that year, and the Clean Air Act in December. That said, however, vigorous and widespread public involvement at a level we could honestly call a movement still took some more time, although Earth Day was a great boost.

Following the national attention on pollution brought about by Earth Day, an unknowing housewife named Lois Gibbs was among the first grassroots citizens to take up the environmental banner, reluctantly, and make history. She lived in Niagara Falls, New York and had became concerned about the health of her community. Many children and adults were showing unusually high rates of headaches, nausea and other medical problems. Some neighbors noticed smelly liquid oozing into their basements, and school children would occasionally sink knee-deep into holes filled with smelly sludge. Apparently her neighborhood and its school were built directly on top of an abandoned hazardous waste site and would soon become a pair of household words - Love Canal. Lois quickly found herself a lone voice in the wilderness, arguing for action, clean up and some sort of compensation for herself and her neighbors for their ruined property values. She fought for more than a decade for action before she and her neighbors finally got it. The state bought their land, moved them out and started cleaning up the dump after nearly 14 years of litigation.

That now famous hazardous waste site was actually an abandoned, filled-in canal named after William T. Love, who dug it in 1893 to provide water for a "model city" he was planning. However, times changed and he ran out of money before completing his canal and city. Hooker Chemical bought the land and dumped its waste into the canal until 1953, when the company covered the canal with clay and sod and abandoned the site. By 1960, the Love Canal subdivision, including the school, had sprung up on the site. During the next decade, seepage from 22,000 tons of goop, hidden by the clay and dirt cover, eventually began to find its way into neighbors' homes and bodies, making Love Canal the single worst environmental disaster in our nation.

We could call Lois Gibbs the mother of the modern environmental movement. It was a grassroots phenomenon joined by other mothers, students, fathers and even grandmothers and grandfathers who quickly learned from Lois that if they wanted clean air, water or land in their neighborhoods, they would have to learn two things and learn them fast: the sources of pollution, and how the political system worked to correct them. The clear, simple goals of finding pollution and stopping it were appealing. Moms, dads, grandmoms, granddads and children picketed the legislature and the polluters with a clear demand for clean air, land and water. And the movement's impact was great. It birthed a new clean air act, new industrial pollution reporting requirements and the Superfund act, and candidates everywhere were taking up the cause. Nearly everyone could feel like a part of the process started by those first middle class citizens. Their movement continued for nearly two decades, successfully placing clean-up of the environment on the desk of the President and most of Congress.

The new "environmentalists" came in every shape and size and developed as many strategies. Living room meetings, community center rallies and church sermons all became focal points for the movement. As soon as one member learned the hazards of a particular chemical, the nuances of a state regulation or the details of an incinerator permit application, he or she passed it on. Phone banks and mail lists were the primary means of communication and organization. In this way, an entire body - perhaps an entire generation - became well educated about pollution sources as well as solutions.

But in Louisiana and elsewhere, this new life on the front line of the environmental battleground, filled with public hearings, protests, rallying cries, tedious research and governmentally slow progress was not easy. Les Ann Kirkland, a jewelry maker-turned-environmentalist living in the riverbank town of Plaquemine, Louisiana, once told me, "I feel like my life's been stolen from me." In order for her to stand up to the large chemical companies looming just outside of town and to ply through layer after layer of governmental bureaucracy, she was forced to become an expert on local pollution. Rather than spending her days making jewelry in the shade of the surrounding live oaks and stately old Victorian homes, she began to educate herself on everything from the political process of waste discharge permitting, to the makeup of chemical ingredients, products and wastes most of us had never heard of, Ike PCPs, PVCs and PETA. She became a living, breathing expert on environmental issues the same way Abraham Lincoln learned the law – on her own time with borrowed books. She called herself a "frontliner," but lamented the fact the movement robbed her of all pleasure reading.

"I used to have time to sit and read, really read. Read books," she told me, "All I ever read now is technical reports. I feel like I don't have time to do anything else."

But she needed help wading through the mound of technical manuals, government documents and legal briefs facing her and her neighbors, and was convinced anyone could jump in. So, like all good grassroots activists, she tried to recruit.

"If they (other citizens) just take up a little slack, the people like me won't have their life blacked out for 5 years."

Still, she often pursued polluters independently and regularly, becoming familiar and comfortable in the local news media and on national TV. And typical of many grassroots activists, environmental or otherwise, Les Ann was driven by more than just a need for clean air and water.

"I still have people who helped me who I need to pay back. That's why I'm here, to pay back. Maybe it (personal debt as a sin) is just one of those old Catholic things, but I have to do it."

In Rochester, New York parents were concerned about the smell in their children's school. Those moms and dads discovered the nearby Eastman Kodak plant had been leaking and suddenly found themselves in a face-off against big business and state government. In that case, the parents also discovered what many grassroots activists have discovered and re-discovered through the years - their government officials may not automatically be on their side. Those parents fought a long, hard battle to get their state officials to recognize the extent of the leak and its effects and to take appropriate actions. At one point, the students themselves got caught in a tug of war. Parents kept their children out of the school until it was cleaned up to their satisfaction, while state officials continued to announce the school was safe, and the children should go back to class.

Lying on the banks of the Mississippi River between Baton Rouge and New Orleans, is the small town of St. Gabriel. Residents there have long since grown used to the almost surreal sight of farmers plowing, cutting hay or baling against the backdrop of the huge, shimmering, steaming towers, tanks and piping of several major chemical plants. One day, a pharmacist there named Kay Gaudet started charting the number of miscarriages in her community and concluded the rate was too high. She eventually persuaded the state health department to oversee a study of the problem conducted by nearby Tulane University.

Grandmother Miriam Price from near Morgan City, Louisiana joined the movement when her granddaughter, along with several other children, contracted neuroblastoma, a rare cancer of the brain. Miriam and others believed the cancer was caused by carcinogens in the environment. She and her neighbors concluded the number of neuroblastoma cases in their community was simply too high for such a small population and began to educate themselves on the disease and on local polluters. Their primary target became Marine Shale Processors, a local hazardous waste incinerator they blamed for the bulk of the area's pollution. Those few citizens were able to focus a spotlight for years on the company, and Marine Shale was eventually fined heavily by the EPA. The company is now closed and looking for a buyer for its assets.

Mary McCastle is a little woman who lives in Alsen, Louisiana, a community just down the road from Rollins Environmental Services, another hazardous waste incinerator with a track record of pollution. McCastle became a local leader in the movement because she simply got scared of the air she and her neighbors were breathing. Through years of pressure, community meetings, church meetings, picket lines and calls to the media, she and her neighbors succeeded in getting a "Closed" sign hung on Rollins' fence, although at the time corporate execs stated the site was only temporarily inactive due to a downswing in business.

Some citizens became semi-professionals within the environmental movement, leading, planning and recruiting others to join the cause. Marylee Orr, a soft spoken woman from Baton Rouge, Louisiana, is a perfect example. Marylee is a self-taught, volunteer-turned-pro who now leads what may be the last grass roots organization of significance in the state of Louisiana, the Louisiana Environmental Action Network, better known as LEAN. Her job isn't just to go out and challenge the polluters, which she does expertly, but to coordinate the challenges of all the other organizations which have joined the network, provide technical know-how and leadership. Her list of associated organizations has the authentic ring of true, grassroots America: SLAP - South Louisianans Against Pollution, C-FACE - Citizens for a Clean Environment, Ascension Parish Residents Against Toxic Pollution, Friends of the Environment and C.A.N.T. - Citizens against Nuclear Trash. Orr's is truly a full-time, educational, motivational and outreach position, relying on contributions and fund raisers to stay afloat. Marylee and her volunteers work out of a bare bones office in Baton Rouge where they house a small but valuable library of technical manuals and background information. LEAN produces summaries of actions by state and local lawmakers, a newsletter filled with advice and examples of success and even "how-to" pamphlets and videos for her member organizations to use in planning their own protests or political strategy. Personally, I think Marylee, her volunteers and the LEAN office are the living definition of a grassroots effort.

Finally, an example of a lay person-turned-expert in a most impressive way is Pam Kaster, a self-described housewife from Baton Rouge who enjoys raising horses. During the last decade, Pam joined and led Citizen's for a

Clean Environment, a Baton Rouge environmental organization. She has become a very recognizable figure in dozens of public hearings on one environmental issue or another and was appointed to her city's ozone task force, local emergency planning committee and various governor's advisory councils, while also speaking internationally on the environmental issues as a concerned lay person. Her level-headedness, clarity, assertive questioning and impressive knowledge of the issues have made her a sought after member of various groups and working committees. Pretty impressive for a petite, soft-spoken, self-described "housewife."

Each of these characters are very real and were very important members of the group of lay researchers and volunteers who fought, dragged and carried the environmental issue from grassroots America to the White House and Congress. They had many successes and many failures, but set a national example for how everyday people could understand the complex issues of a clean environment and affect change at the highest levels.

One example of how everyday people like these learned the issues and turned an industry-wide practice on its ear in Louisiana and elsewhere was the notion of "risk." Early on in the movement, environmentalists made a significant discovery. Government permitting agencies had a health requirement which industrial companies had to meet before receiving permits to discharge emissions into the environment. Those companies had to determine the health risk their discharges posed to nearby residents. A company proving their emissions presented less than one-in-a-million chance of neighbors becoming ill or contracting cancer could walk away with a permit. Well, one-in-a-million may sound pretty good to you and me, because those lopsided odds appear to be very much in our favor, but the new environmentalists didn't see it that way. They spotted a flaw and deftly used phone banks, living room meetings and community center rallies to get the word out. What they discovered was that all kinds of new industrial equipment, monitors, techniques and know-how existed to reduce cancer risk from pollution even more than the standards allowed. But so long as a company could meet the one-in-a-million standard (or whatever risk the permitting agencies required), it did not have to spend money on the latest and greatest pollution reducing stuff.

Environmentalists said "nuts" to that. They began calling for industry to use the best available technology in all construction and permit procedures, not just those where initial risk was high. Through this revelation of risk, and the subsequent pressure on lawmakers and industry, grassroots grannies actually managed to get big industry to stop using low risk as a design criteria (as though very low risk of cancer was acceptable) and start using the best available technology and practices. Now, many permitting agencies require industry to prove they will install the "best available technology," also called " MACT - Maximum Achievable Control Technology," and use the "best available practices" before granting a permit to operate.

The grassroots movement also tackled the notion of "waste streams," a term applied to the very real and very definable streams of waste emanating from any industrial site, whether the waste was to the air, the water or to nearby landfills. The reality was that industrial site managers knew to a very accurate degree the composition and amount of their waste streams. Many lay people in this country began to gather information and techniques from their counterparts in Europe, who had been identifying waste streams for several years. It became apparent that the old adage we use describing "one man's trash" as "another man's treasure" applied very well to industrial waste also. European environmentalists made the argument that if a company knew what was in its waste and produced a steady quantity of it, that "waste stream" could actually become a stream of raw material for someone else. Just as we might sort through our garbage at home to identify and collect recyclables, industry could and should recycle its waste. Grassroots organizations in this country heard that in some parts of Europe, companies applying for operating permits had to identify specific companies (even competitors) who were willing and able to use any newly created waste stream as a raw material. Local environmentalists brought that argument home to pressure industry to do the same thing here and eventually government permitting agencies and many companies saw the economic, environmental and public relations wisdom of identifying and using waste streams. Congress created a new law to cover such practices called, appropriately, the Resource Conservation and Recovery Act. The act specifically calls for companies to reduce its waste, re-use its waste or recycle it. Chock up a victory for the little guy! Grassroots volunteers saw their idea end up as law.

The massive Dow Chemical site in Plaquemine, Louisiana became an American example of how to identify and use a waste stream. One of the plants in the sprawling Dow complex manufactured antifreeze. The vessels and pipes of the plant which carried out the high temperature chemical reactions required cooling water, but during the process of cooling, the water became tainted with ethylene glycol, the main ingredient of anti-freeze. Dow's old procedure was simply to send the used water to a water treatment system to dispose of the ethylene glycol and water. But by 1990, following the national trend to recycle, Dow developed a way to remove the ethylene glycol in the cooling water before it went to waste treatment. Voila! A waste stream was then turned to a raw material stream. All Dow had to do was send the recovered ethylene glycol back into the manufacturing plant. That idea is called "waste reduction" - meaning a company's waste stream was reduced because most of it was converted into a useable product.

Grassroots environmentalists' phone banks, community meetings, newsletters and living room pow-wows spread the idea of waste reduction to moms, dads and school children. All across the land, elementary school students sang, "Reduce, re-use and recycle." I even attended a corporate sponsored, nationally touring play filled with reduce, re-use and recycle songs. It was very effective in bringing the environmental message to the papers and TV. Corporations were also hit with a strong message - reduce, re-use or recycle your own waste. Score another victory for everyday people who took the time and effort to make a change.

The real political strength of the grassroots movement grew from its depth of character or, I should say, its cast of characters. Presidential and Congressional candidates saw impressive, diverse crowds of people gathered on courthouse steps, inside legislative committee rooms and pounding the pavement in front of major chemical corporations. This was the heart of grass roots activism. White collar or blue collar, rich or poor, union or non-union, they all sported the same buttons and carried the same signs. Newspapers and TV soaked it up. Reporters churned out dozens of stories on the personalities alone. And politicians were quick to learn that a sure way to get some votes was to proclaim themselves friends of the environment.

It didn't take long for most candidates to recognize a key characteristic to the grassroots movement - it had no significant downside (to use their own word). Advocating a clean environment was a safe bet for most all constituencies. New federal reporting requirements forced industry to reveal the millions of tons of pollutants it sent into the air, land and water each year. So in a classic public relations move to turn this negative information into a positive story, industry began to acknowledge its pollution, while reporting its efforts to reduce it. It seemed grassroots activists and chemical industry managers might actually agree on the problems and the solutions, and both were seeking public means to show their new color - green. How could a politician go wrong? Green was good, and votes were there to be had from both the corporate world and the grassroots environmentalists. By 1992, nearly every political candidate in the land, including presidential candidate Bill Clinton, started nailing environmental planks to his or her election platform.

It seems clear to me that the diversity and sheer will of the common man (and woman) was the nature and strength of the grassroots environmental movement; the movement that just died. Look around. I challenge you to find as many dedicated lay persons as there once were, leading the charge for a cleaner and healthier environment. Where are the dozens of concerned citizens meeting in living rooms, churches, community centers and back porches? Where are the small groups gathered to hear news and advice from the traveling experts who made their way from town to town to educate and inform? Where are the placards, the stickers, the buttons, the cars and trucks jammed outside the state capitols? I'll tell you. They were rendered obsolete by slick, professional and well-orchestrated campaigns designed to grab money, notoriety and power. Yet as effective as the professional campaigns were in gaining press and action, they eclipsed the grassroots movement, drove out many moms and pops and left only a shadow of a movement. And now that the pros have overgrazed the grassroots environmental movement, they are moving on to new pastures. The movement as we knew it is dead. Read on.

# The Death of the Movement

Should you already disagree with me about the actual passing of the environmental movement, allow me to point out some indicators and a few noteworthy participants in the birth and subsequent death of the movement.

The first Earth Day of April 22, 1970 was conceived and carried out by then U.S. Senator Gaylord Nelson of Wisconsin. He said he got the idea from anti-war "Teach-Ins" which were being held on campuses across the nation. He felt a national environmental teach-in of some sort was in order and decided to organize one. The time was right, the sentiment ripe. Earth Day was a huge success, with an estimated 20 million people (mainly college, high school and grade school students) participating in peaceful demonstrations and clean ups. Unfortunately, To borrow a phrase from today's conservatives, the day ended up being a "feel good" event rather than a significant step toward a cleaner earth. National sentiment quickly shifted from the environment to the ending Vietnam war, then to the energy crisis and later to Glasnost. The environment wasn't really on the national agenda again until the late 1980s.

During his term from 1989 to 1992, following the lead of a reinvigorated movement, President George Bush declared himself the "environmental president." Bush then led the nation to a revised Clean Air Act, a new Energy Policy Act and a revival Earth Day in April, 1990, the 20th anniversary of the first. During the presidential campaign of 1992, Democratic candidate William Jefferson "Bill" Clinton campaigned heavily on the environmental issue. Later, during his first term, he invited a group of residents from Louisiana to visit Washington, D.C. for a bit of public relations on the issue of clean air. Clinton, on national TV, took audience questions about the environment. On that day, Clinton listened to a small boy claim that pollution killed his brother then ask what he, the President, would do about it. Clinton responded knowingly, acknowledging the boy did happen to live in an area known as "Cancer Alley" and that he, as President of the United States, would do something about pollution. Bingo! A simple, little movement just landed firmly on the White House agenda. What a huge success in a short period of time. A grassroots movement couldn't have gotten any greater exposure than that, except maybe from Oprah Winfrey.

Oprah had an even greater audience than President Clinton when she broadcast her show from Plaquemine, Louisiana to tell the world she was concerned about the unusually high rate of cancers among the poor, black people living in the shadow of the huge chemical plants there. The city auditorium was packed with grassroots grannies, moms and dads who had a chance to ask questions and offer opinions. It was a perfect blend of emotion and hype that created more than one heated exchange between industrialists and environmentalists. Following her show, Oprah held a press conference to tell the nation that chemical companies should develop a "moral consciousness." Wow! Talk about your successful movement. All the work of those common folk was paying off again. They were able to dominate an hour of national TV in a prime spot, while also gaining plenty of news coverage in the days leading up to the show, during the show and for weeks afterward as other media joined in. Public relations professionals across the country with fat salaries, years of expertise and corporate bucks to spend could not have done any better.

By the way, both Oprah and Clinton were victims and perpetuators of what I call data slinging. I'll define data slinging this way: it is the careless use of data to draw misleading conclusions, especially when the data is misused, misunderstood, poor or lacking altogether. Environmental extremists were good at it and caused both Clinton and Oprah to misrepresent the potential harm in "Cancer Alley." Nevertheless, the grassroots movement latched onto the "Cancer Alley" slogan and used it to great advantage until data slinging caught up with them, eventually contributing to the downfall of the movement. I'll write more about that later, but for now I argue that misinformation or not, few - perhaps no - grass roots movements have climbed so high and so fast while attracting so much direct attention from political figures and national talk show hosts. Although the first Earth Day of 1970 may have marked a beginning more than a decade earlier, the environmental movement became a distinct part of the 1980s when environmentalists forged their huge national presence. The results were just as impressive on the local congressional trails. In 1992 and 1994, congressional and local campaign platforms didn't just contain environmental planks, they were framed, walled and trimmed with them too. Environmentalism had become a household word.

So where were these environmental issues a mere two years later? Campaigns for a cleaner environment apparently gave way to the more common matters of taxes, health care, welfare, jobs and the deficit. It was as though the environment had become suddenly and completely clean. Almost none of the major campaigns and few of local and state ones touched environmental issues. In September of 1995, the Louisiana Environmental Action Network held a forum for Louisiana candidates for U.S. Senate and the governor's office. According to LEAN, it was the sole forum to focus on environmental issues out of the 150 forums held that campaign season. That alone was a good indicator the environmental movement was dying.

Another indicator was in daily journalism, where shrinking news coverage reflected the drop in citizen concern about environmental issues. In 1990, a new professional association sprang forth from the halls of newspapers and television station news rooms, calling itself the Society of Environmental Journalists (SEJ). One of the senior environmental reporters in the land, Jim Detjen of the Philadelphia Inquirer, was SEJ's first president and one of its founders. The SEJ's purpose was to allow a growing field of specialized environmental reporters to hone their craft by learning more about the technical data that was available to track and understand pollution, by swapping stories with their peers and by gaining insight to gathering news about the environment. The organization grew quickly as more and more news organizations began placing reporters on the environmental beat. That's significant. News organizations are generally reluctant to put reporters on single beats, because that move leaves fewer people to cover all the other daily stories. However, the depth and breadth of the movement, the cast of characters involved and the high profile responses of our political leaders created a lot of newsworthy items. There were clashes between grandmothers and giant corporations, colorful, hippie-like members of Greenpeace donning scuba suits and zipping around in inflatable boats to "plug-up" polluters of the Mississippi River, mean spirited plant managers threatening to fight them off with green-colored ax handles, and lots and lots of school children learning to "re-use, reduce and recycle." Talk about news! Ink was flowing and cameras were rolling.

    Many reporters who were used to general assignments soon found themselves with plenty to do covering just the environmental movement. Many news organizations, to their credit, began to seek and hire reporters with technical backgrounds, college degrees or advanced degrees in the sciences to cover the growing complexity of data on the quality of our air, land and water. That was how I got my start as an environmental reporter at WBRZ-TV in Baton Rouge. Although I may have been an unlikely candidate for general assignment, my science degrees and an understanding of techno speak must have impressed my new boss John Spain, news director at WBRZ. I had applied for a job as science reporter and weather forecaster, but was soon drafted to cover the environment. It was a great ride while it lasted.

By late fall of 1997, the SEJ had become an organization representing a trade in distress. WBRZ and other news organizations were devoting less time and space to the environmental beat, former star environmental reporters had moved on to other interests, and the funding and future of the SEJ itself was becoming questionable. Whereas a few years prior, the SEJ could entice notable, national level keynote speakers, hopefuls like EPA administrator Carol Browning and her Canadian and Mexican counterparts turned down SEJ invitations for its annual meeting in October of 1997.

Why? I believe a lack of clear direction in the movement itself was reflected in the shrinking pool of those who reported on it. The environmental beat was becoming more splintered and less defined and suffered a credibility problem. It simply was not yielding much news, and reporters were dropping off the beat and out of SEJ.

I'll spell out the exact credibility problems later, but for now let me say this. During my years on the environmental beat, I learned several things about the movement and the people involved in it. First, the movement's lifeblood - and nervous system - were the many everyday citizens who got involved. This huge pool of conservatives and moderates made phone calls, hosted living room meetings, attended public hearings, and even walked the picket line. Most of these simply wanted to stop intentional pollution by what they viewed as uncaring or uninformed corporations. By uninformed, I mean the grassroots folks believed the company management simply didn't know the communities their companies did business in. The grassroots members believed the managers didn't know the people, the problems or the perceptions citizens had toward them or their companies. For this reason, most of the energy in the early environmental movement was focused on a strong message to companies: get to know your neighbors and neighborhoods; listen to them and learn from them. Unfortunately, many chemical companies still had what I call a 50's approach to public relations. Company "spokesmen" would stop reporters and citizens at the front gate, believing secrecy was the best policy. But grassroots members didn't stand long for that kind of stonewalling, and so they appealed to the companies time after time to talk.

To me, the desire of moderates to sit down and talk calmly with industry executives was a common sense approach to alleviating misunderstandings on both sides of the corporate fence. To be sure, citizens didn't always sit across the table from plant managers who were open and honest, but the managers didn't always find sincere, reasonable people at their gates, either. But, that said, I believe the greatest strides occurred when both sides did indeed try to understand each other. The reality I saw many times, though, made the process of reasonable talk very difficult for the grassroots soldiers who wanted to try. Many environmentalists demanded extreme changes in industry, society and even personal lifestyle for the sake of the environment, so I'll describe them as such: extremist. By this I mean they would nail chainsaw-stopping spikes into trees, block off chemical company discharge pipes, hang banners on bridges and tie up traffic rather than sit down and talk. Even less extreme individuals in the movement often presented unrealistic demands of industry and of society. Believe it or not some people out there would like to see, for example, a ban on chlorinated drinking water, even though that was one of the greatest public health strides our country has ever made. Goodbye chlorine; hello bacteria. Some folks would also shut down plants and do away with products like rubber for tennis shoes, chlorine for medication, fuel for cars, and plastics for just about everything including the word processors from which they churned out their flyers and leaflets.

These extremists were in the minority of the movement, but were surprisingly successful in setting much of the agenda by choosing the tactics and setting the goals of the movement. They were like overzealous country evangelists, swooping into town, setting up a tent, preaching for three days straight to win truckloads of converts, collecting the offering, and then leaving town while the newly converted wandered around in a state of slight confusion about what to do with their newfound salvation. I assume others on the environmental beat observed the same extremism, which is why I am disappointed at my colleagues (and myself) for not reporting more about it. Maybe more in-depth reporting of the cast of characters, the agendas and the motives involved in the environmental movement would have kept the everyday public's interest alive in the issues and in the people involved.

But extremists, sophisticated agencies and professional organizations disillusioned grandmothers, children and lay people who disagreed with political agendas that had reached far beyond cleaning up the neighborhood. Professional environmentalists left locals behind and moved to the highest, national levels to argue about the criteria industry used in selecting sites of operations, hiring workers and swapping pollution credits. Their battle cry was environmental equity, and it left the conservative, middle-class grassroots members behind, wondering what happened to their simple movement.

As a news reporter and as a news reader I can tell you this. As soon as leaders of the movement started arguing complex policy by tossing around big, vague words like environmental equity and social responsibility, and when extremists started spitting out too many unfounded accusations, like chemicals are killing us, everyday citizens tuned out. Like reporters, citizens wanted action and facts, which meant news, and they weren't getting any. The long, convoluted process of formulating policy was rarely news, and poorly defined terms and mean-spirited comments only clouded the issues, particularly when they involved groups with unclear environmental motives like trade unions, civil rights groups, legal aid groups and so on. By the time agencies and boards could get around to putting policy into action, which might have been news, the majority of the moderates and the media had left the room. According to former Society of Environmental Journalists president Jim Detjen, who is now the Knight professor of environmental journalism, 44 percent of reporters he polled said they were spending less time on environment stories by 1996. That news was echoed in a study of news content by the Center for Media and Public Affairs (CMPA), which revealed by late 1997, network news had cut its environmental coverage by two-thirds from its 1990 levels. That meant fewer people were involved in the environmental movement and therefore were creating fewer stories. Some people may tell you that the news media cut back on the environmental beat because the new clean air act actually worked, or that industry is now polluting less, or there just aren't many disputes left, meaning the environmental movement was successful. Hardly. Reporters cover fewer stories because the movement was stolen from the grassroots grannies and taken to the halls of academia, the offices of Greenpeace, the AFL-CIO, the NAACP, the ACLU and other worldwide organizations which rushed into the grassroots environmental movement and wound up competing for membership and money to push their own agendas. In effect, the national organizations made the local environmental beat a non-news beat, replacing clean air and water issues with difficult to follow issues on civil rights, union benefits and even a new social order. Journalists were moving to richer grounds for hard news and

## Goodbye Green                                      Duncan

citizens were dropping out of the movement and shifting their volunteer time and talent to things more meaningful to their personal lives.

    Therefore, the most important death sign of the environmental movement was also the most simple - lack of participation. Just as fewer reporters covered the environmental beat, fewer citizens made news for reporters to cover. Some were pushed out or disillusioned by the extremists, others lost their way during the confusing policy debates, while some simply got tired of shouting, wanting instead to sit down and work out problems. I'd like to give you hard figures on this, but I'm not sure anyone was counting. I can give you my personal observations from several years on the beat, though. At one time I estimated several thousand people at an environmental rally at Louisiana's capitol, and that was just one of several rallies held there in 1988. During that same year, my colleagues and I could rely on a different group to show up at the capitol nearly every day during the legislative session, sporting pins or wearing tee-shirts and carrying signs demanding action on any number of environmental issues: stop the construction of a landfill, outlaw hazardous waste disposal wells, make industry tell us more about the stuff it spews out, install more warning devices, make getting air, water and land disposal permits harder, clean up Lake Calcasieu, stop dredging for shells in Lake Pontchartrain, do something about saltwater in the estuaries, shut down Marine Shale's incinerator or Rollins' incinerator (or whoever) and on and on. Also, because the revival Earth Day in 1990, and its next few annual celebrations, occurred right in the middle of our state legislative session, reporters and citizens alike could spend entire days on the steps of the capitol listening to enviro-speak, watching enviro-skits, learning about enviro-issues and passing the time with enviro-moms and grassroots grannies. The crowd cheered when "recycle" man showed up at the Capitol one year, a twenty foot tall man constructed entirely of recyclable materials dangling from a pole held up by local enviro-kids. And don't forget the stilted stork man, a man in a nine-foot tall stork costume walking around the crowd, reminding everyone of pollution's harm to wildlife.

As I recall, that stork man first appeared in the Great Toxic March from Baton Rouge to New Orleans in 1988 but quickly faded from the scene. The reason, as I see it, was that the Great Toxic March marked the beginning of the end of the grassroots movement. Certainly it was the peak of the movement in Louisiana, but the slide downhill began almost immediately after. By 1997 the crowds, the enviro-songs and enviro-skits were gone, the enviro-moms became soccer moms and the grassroots grannies went back home. If my observations are correct, the entire legislative session in Louisiana that year passed without a single grass roots environmental turnout of any significance. Yes, the local organizations had a small presence, but gone were the throngs, and gone were the TV cameras and newspapers. The two major TV stations in Baton Rouge (ABC affiliate WBRZ and CBS affiliate WAFB) no longer had anyone described as an environmental reporter, and the daily newspaper, The Advocate had assigned its long-time environmental reporter to a much more general, suburban beat.

Hmmmm. The symptoms were compelling: weak pulse, few break outs of demonstrations, news coverage diminishing and less people involved. The movement's last gasps were audible both locally and nationally.

Of course, the great masses who left the movement were the same who swelled its ranks earlier - conservatives and moderates. The simple, effective movement we saw bloom and mature in the 1980s had become weedy and rangy. What was once a strong movement, joined by Americans from all walks of life seeking very specific environmental results, had now petered out to a vocal crowd here and there arguing for rather vague notions of civil justice, usually directed toward a single company or industrial site. But hold on, the next chapter is the best part. Now that you've seen the evidence that the grassroots movement really is dead, you'll catch a glimpse of the intrigue and manipulations of the really big environmental, labor and civil rights organizations as they swooped in to join the grass roots movement - just before strangling it to death.

# The Price of Success - How the Movement Died

The late Congressman Tip O'Neill once quoted his father as saying "All politics is local." Likewise, the heart of the environmental cause was the notion that cleaning up the whole earth began with hometown citizens working on hometown problems. This was the core of the environmental movement; a movement literally and figuratively cultivated from the grassroots up. Citizens wanted clean air, water and land and were gathering outside the gates of their local polluters to demand changes. Those citizens found that rather than working toward a complex, vague and perhaps unproductive national strategy like a new clean air bill, they could gain greater satisfaction and results by joining groups which were picketing their hometown landfills, chemical companies and government officials. Citizens striving nationwide to correct their local environmental wrongs added up to a national movement.

This collection of everyday citizens, learning from one another, encouraging one another and prompting change in their communities created a network and method that could be the blueprint for any other successful, national grassroots campaign. Local organizations needed no orders handed down from a national office, they needed only direction and technical assistance. For this, either self-appointed or chosen leaders appeared on the scene, making themselves available to every local organization nationwide to swap success stories, strategies and information. Leaders formed workshops, seminars, round-table discussions, environmental festivals and more Earth Days. Because the goal of the movement was clear from the beginning, this national swapfest of information and methods was ideal. This cheap but effective strategy allowed every hometown citizens' group to pull in roughly the same direction at about the same speed. Not even Ross Perot could accomplish the same thing with his failed "United We Stand" grassroots campaign for President.

The movement, when it was alive, was actually pretty neat stuff. Moms, dads, grandmoms and granddads who were tired of looking at polluted lakes, factory smokestacks and smelly landfills found themselves on picket lines and in mayors,' governors,' and state legislators' offices, urging them to clean up their towns. Politicians, factory managers and the media had to take notice. The new comers to activism were hard to ignore. Here was a well-organized group of gray-haired people, school children, young moms and white collar dads. They created an extremely productive public relations campaign which captured the imagination as well as the air time, news space and cooperation of their local media.

When news organizations polled citizens about their worries, the environment was right up there near crime and losing good jobs. So news organizations began to cover more environmental stories, and more and more TV viewers and newspaper readers saw folks just like themselves in the news trying to make a difference. Guess what. The readers and viewers soon joined in, starting a predictable snowball effect. Grassroots groups would call a news conference in an appropriate place like the steps of the capitol or in front of a large corporation or at a polluted lake. The media would show up. Then the group would announce a rally, and the media would kindly show up and report that too, sometimes more than once. Moms and dads and schoolchildren would then flock to the rally they just heard about on the news, and a crowd would grow. The media would show up again - to the rally it just advertised - and would report on the event and the large crowd. Those rally goers would end up on TV and in the newspaper. That meant even more viewers would see people just like themselves on TV trying to make a difference, and those new viewers would join the next announced rally. That, friends, was classic public relations and a perfect example of a partnership in which the media helped create its own stories. Politicians and public agencies could not ignore the newly created crowds. The environmental movement was of the common man, uniting entire communities, swelling audiences for speeches and delivering votes. More importantly to the arguments of this book, though, the huge crowds really meant two things: power and money. And folks, when you start throwing around power and money, plenty of people will fight to grab it, and fight they did.

Two battles for control of the movement began almost at the same time. One was from those already in the movement, the other from those outside who desperately wanted in. Let's begin with the battle from within the movement. The simplest description of it would be a feud between the moderate members and the extremists. Moderate enviros (my term) were those members of the movement who believed that most solutions to pollution problems would stem from cooperation and collaboration with those responsible for causing pollution, namely the large chemical corporations. Also, they believed that curing pollution was everyone's job, themselves included. These were the folks who fought long and hard for education, training and action in recycling, automobile emission controls, car pooling and environmentally safe packaging, and a reduction in what I'll call "personal pollution," things like household hazards, paints, poisons and plastics. This part of the moderate environmental movement was highly successful in suggesting that we, the consumers, were all responsible for pollution, and we all had to spend time and money to solve it. We listened, gladly paid higher prices for products and altered our lifestyles to accept each change. You may remember events in your own town with names like Household Hazardous Waste Day or Recycling day when a local chemical or garbage company set up a tent or brought trucks to collect all your household hazards or recyclable materials. City, county and parish governing councils across the land rushed to establish community-wide recycling programs. News stories, education booths and even government programs sprang up to educate us and stop us from changing our car oil and dumping it down the drain, or from using our garden hose to rinse out paint brushes in the yard. We learned to recycle our tires, car batteries and even refrigerators. We were even taught to distinguish between good ozone and bad ozone: ozone was good when it was high up in our atmosphere, forming a protective layer against the sun's radiation; ozone was bad when it was down here on the ground, making it hard to breath. To protect that thing called the ozone layer (good ozone), the entire air conditioning industry switched from one type of freon to another so leaks wouldn't be so damaging, and a maze of pump bottles appeared where hair spray cans used to be to eliminate ozone destroying fluorocarbons. And because car and lawnmower exhaust and outdoor grill smoke helped form ground hugging ozone (bad ozone), car pool lanes appeared on the freeways, and residents in some cities had to stop outdoor grilling and mowing lawns on high ozone days. Also, new fangled pump nozzles with funny looking, elephant-like snouts appeared in gas stations designed to prevent gasoline vapors from escaping (another ingredient to bad ozone formation). We all pitched in because we learned we were the problem.

 Nuts to that, said the extremists, the other side in the environmental movement. In their eyes, the chief causes of pollution were those huge smokestacks, leaky landfills, waste injection wells and hazardous waste incinerators dotting the landscape. Their point was that industry caused pollution, and industry must pay to stop it. They also claimed industry did not own up to its pollution, so managers could not be trusted in any kind of collaborative or cooperative effort. In fact, though industrial managers began extending invitations to tour their plants, the extremists rejected the invitations, referring to them as thinly veiled public relations stunts which would reveal nothing. So vehement were they about this, they could not and would not entertain the notion of actually sitting down with industrial leaders and working out a solution together. After all, it was hard to call industrial leaders polluters, liars, poisoners and sinners at public hearings, and then sit down to have a friendly chat with them about ozone reduction.

Wow! That was quite a schism in such a nice little ol' grassroots movement. The moderate vs. extremist fracture first appeared to the public in the days leading up to the Earth Day revival in the spring of 1990. A preliminary event called Earth Fest '90 was organized by grassroots moderates and planned for a weekend in March at the Audubon Zoo in New Orleans. Earth Fest organizers had requested and received substantial donations from private sources, and things seemed to be going well until extremists learned Shell Oil had contributed money. (A note here. At the time of Earth Fest, New Orleans was a center for oil drilling, and most major oil companies had a large presence. Any fund raisers would be foolish to ignore the money those companies were willing to give to good community programs.)

"No Way!" said the extremists, "Earth Fest should not be tainted with money from polluters like Shell Oil or any of those other big chemical companies and refineries. They were the problem. They couldn't possibly be part of the solution."

So, what to do? As part of a continuing effort to become "green," a New Orleans civil rights group called The Gulf Coast Tenants Association filed suit to exclude Shell from

**Goodbye Green**  **Duncan**

Earth Fest '90 even though moderates wanted the company's help. Thus the first crack in the great green movement became visible to the public. Despite the hoopla, Shell remained as a participant in Earth Fest. Greenpeace, however, did drop out of the following Earth Day.

Shell Oil, perhaps, became the scapegoat for what extremists viewed as infiltration of their movement by big business. To emphasize their point, some extreme enviros held an Anti-Earth day for themselves. This small group of seasoned activists from the Baton Rouge area met in the living room of one of its members during the Earth Day festivities and "took the day off," as they told me. They claimed they had worked day in and day out on environmental issues for most of their recent years, and the Earth Day revival, once tainted by big business, was the perfect time step to out of the picture in protest.

On that day, while school children, parents, grandparents and industrialists alike recognized the importance of a clean earth in rallies, fairs, booths and speeches nationwide (an event many, many people called a huge success for getting the green message out), this handful of radicals dropped out in protest, clearly foretelling the nature of the movement's future. The local papers and TV missed the story. Instead, seeking the quick, easy and obvious, local media reported on the fact that, ironically, litter was a problem at Earth Day. The crowds were bigger than expected, and planners just didn't put out enough trash cans. Litter stacked up, fell over and scattered across the site. Okay media, forget the environmental drama playing out down the road, just report the litter.

That day's parting of environmental philosophy alone would not have killed the movement. Any large movement will naturally have differing opinions among its members. For example, as successful as the civil rights movement was, it was marked by huge differences in philosophy, peaceful protest vs. militant reactionism to name one. However, the early appearance in the environmental movement of the same philosophical difference created the crack necessary for larger, more powerful and more extreme groups to squeeze in and begin to shape the movement, turning it from a local grassroots effort to a well-orchestrated, professional, national one. That's a problem, because most grassroots participants, by definition, are local residents with mainstream ideals. Indeed, true grassroots movements are by and of the common man - they are an "everyman" movement, and that is exactly how they become so powerful. But the only way to include every man (or woman) is to stake out a moderate position. Extreme positions tend to thin the ranks.

Early environmental positions and tactics were moderate and local – peaceful protests, media events, participation in public hearings, letters to legislators and the like. This moderation swelled the ranks of grassroots organizations, but it also made them targets of large, national organizations wanting to capitalize on the trend to go green. This second battle for control of the movement - from people and organizations outside the grassroots movement - appeared to be started by extremists within the local organizations. From where I sat, the extreme element captured the lead on several fronts shortly after entering the movement, and for two good reasons. First, its positions seemed clear, firm and apparently supported by data rounded up through the workshop and seminar grapevine. This gave the extremists an air of expertise and wisdom. Second, extremists were strident and visible, and both of those qualities attracted the media, netting huge news coverage. Many moderates began seeking guidance from their seemingly more experienced, better prepared, extreme colleagues. Like eagles after prey, those extremists quickly spotted their chance to gain some really big clout and control. They called in the big boys – national trade unions, Greenpeace, the NAACP and Citizen Action to name a few. These so called "national experts" started showing up in local grassroots living rooms and meeting halls, coaching, cajoling and offering technical assistance and specific information on the polluters in their area. Paid by national organizations like workers' rights groups, trade unions and civil right groups, these professional environmentalists flew in to groom regional and local activists who soon made environmentalism their career. These new enviro-pros with the familiar faces were soon well equipped with data, training and expertise to offer their community volunteers the "inside story" on polluters in their area and what really went on inside the gates. The pros also offered a seemingly broad, national and international perspective on decisions world corporate leaders were making concerning pollution and how those decisions impacted chemical plants and local citizens from Germany to Massachusetts to Louisiana.

The original grassroots leaders and participants - the locals - were thankful for the organizational help, technical assistance and publicity these national and professional experts could provide. Plus the national groups had money to stage the necessary events to get attention. The trouble was that their money came with strings attached; those larger, national groups wanted to control the agendas, schedule the rallies and shape "the message" of the movement to the media. (note: in public relations, "the message" is everything. It is the main point you wish to make either in speech, writing, action or demonstration. It involves statements as well as responses to media questions or critic's statements, so whoever is shaping or crafting the message is controlling the public relations.) This rise of the professional environmentalist was yet another one of the early signs of the movement's demise.

What happened was simple. Grassroots leaders were hungry for help and more experienced guidance. National organizations and liberal agencies sensed it like a bloodhound hot on the trail. In no time, Greenpeace, the

AFL-CIO, NAACP, ACLU, Citizen Action, the US Department of Justice, the Presidential Commission on Environmental Justice and other national groups swooped in to establish a community presence. Awaiting them, and ripe for the plucking, were local environmental manpower, membership and money along with the precious media attention locals could muster. Whoa! A simple, effective grass roots movement was fading fast. National trade unions, civil rights groups and radical environmentalists had begun to leach power, influence, support, money and attention away from the local groups to selfishly further their own social and political agendas.

Unions saw the local environmental membership numbers and envisioned more clout at the bargaining table. Civil rights groups, having lost their focus, saw industrial pollution as a great, unifying minority issue and made their own grab for local environmentalists and their money. Traditional conservation groups, having fallen behind the environmental movement, got greener by re-making themselves to be champions of clean-up rather than champions of preservation. Even worse, ultra-radical groups like Earth First! saw the popularity of the grassroots environmental movement and used it to legitimize their calls for a new socialism where "the people" (translated to mean the government) would set the agenda for business, development and lifestyle through law and regulation. This rush for members and money was what killed the grassroots environmental movement. It wasn't the first time greed and hunger for power killed a good thing.

# Labor Goes Green

The reason for labor's rush into the environmental movement wasn't clear to me at first. In fact, it seemed counterproductive. Although workers clearly had a right to clean air on the work site and to be protected from hazardous chemicals, these were not exactly environmental issues. They were worker safety concerns, and unions had been well-equipped and motivated to deal with those issues for years. However, rather than push for higher wages and more benefits, unions began to pressure employers to spend millions of dollars to clean up suspected groundwater contamination, to reduce air emissions by another few percentage points, and to spend hundreds of millions of dollars cleaning up old, abandoned sites. At first that just didn't seem like the good old unions I was familiar with. The reality I discovered, though, was that union pressure on employers to clean up the environment had everything to do with money and jobs, but the connection was not straightforward. The media missed it altogether. In fact, I suspect the grassroots grannies missed it, too. They were just glad for the help.

Here's a Louisiana example of how and why unions turned green. On June 15, 1984 managers of the BASF chemical plant in Geismar, LA and representatives of the Oil, Chemical and Atomic Workers union walked out of contract negotiations without renewing. Shortly after, management announced it was laying off 370 union members. "Foul!" cried the union, "These are not layoffs. this is a lock out." That was no small news story. BASF was the largest chemical company in Germany and the second largest in the world. It's Geismar facility was the largest of its 80 U.S. plants. The facility sprawled over more than 2,300 acres along the Mississippi River and employed nearly four hundred workers. Although management called them layoffs, the union claimed a layoff of nearly all the workers on site was nothing less than a lock-out and began that day to get its members re-hired. The effort would last 4 years.

From the beginning, out of work OCAW members attempted to get the "lock-out" spread throughout the news media in a big way in order to turn public sentiment against BASF and bring public pressure to bear on the re-hiring effort. Naturally, the local union called on its national office for help, which answered by sending a professional labor consultant down to set up an office in Baton Rouge. The move toward environmentalism didn't start immediately, so for two years the union claimed the usual hardships of a lock out: its members were burning up their savings, it was patently unfair to keep workers out who want to be there, it was a money-grabbing effort by a greedy management, and a very complicated and dangerous plant was now in the hands of untrained or poorly trained, inexperienced, temporary workers.

That last argument became the union's first, small step toward environmentalism, because the chief danger union members pointed out was the potential for accidental releases of toxic chemicals due to accidents by the struggling temporary workers. Any large chemical accident would injure more than the workers, nearby citizens were at risk as well. So, a year into the lock-out, workers gathered at the union office and other meeting places to collectively recall all the spills, leaks and releases they had seen. When workers spotted these spills on a map of the BASF site, they quite literally painted a rather large picture of environmental hazards and created a new weapon in the war against BASF - environmentalism. Since the chemicals at BASF were similar to the ones that killed so many in Bhopal, India during that famous disaster, the union erected a billboard warning that BASF could turn into "Bhopal on the Bayou." Wow! Did that ever work. Newspapers began reporting an unusually high rate of accidents and chemical releases during the lock out and the OCAW got the attention it wanted for its cause.

At the same time, the environmental movement was gaining lots of ground and grassroots members. It didn't take long for the OCAW expert to notice and begin coordinating union activities with local environmental radicals, issuing news releases and offering support to local grassroots groups. In 1986, two years into the lock-out, OCAW helped form a group called the Ascension Parish Residents Against Toxic Pollution Think about it. Would a union staff member be uprooted from wherever he was, sent to Baton Rouge to a nice, rented office and be paid a salary from union dues just to help clean up the environment? The answer was clearly "no." His main goal was to get the members re-employed, period. To do that, he chose the tactic of throwing the weight of the state's environmental movement onto the union side of the bargaining table at BASF.

The OCAW office established in Baton Rouge became the center for environmental strategy, a resource for all local environmental groups and a collection point for industrial data. One union member once bragged to me that their files contained the locations of every injection well, the dates of every accidental chemical release and a knowledge of everything the companies in the area were dumping into the air and water. Presumably, this data surfaced at appropriate times during the permit hearings, environmental rallies and contract negotiations with BASF and in contacts with the media. Many of the phone calls I received in the newsroom to tip me off to some sort of industrial pollution during that time came from the locked-out BASF workers or their still-employed counterparts. Those locked out workers turned up the heat on industry considerably. Don't get me wrong, as a reporter and citizen I thought that was a good thing. It forced employers to think carefully about issues larger than working conditions, like how those conditions affected the big picture of accidental chemical releases. It seemed clear to me that unsafe or hazardous working conditions jeopardized more than the workers themselves, because once a harmful chemical spewed into the air, it didn't single out employees and union members, it could drift into nearby homes, too.

On October 21, 1986, OCAW publicly showed its environmental colors by joining the Sierra Club and the Louisiana Environmental Action Network (LEAN) to produce a study of toxins released to the environment by Louisiana industry. The study was derived from data which local industry was required to report to the state's Department of Environmental Quality, but which was mostly unknown to the public. The study brought lots of press attention to the environment, the local environmental movement and to the OCAW.

In June of 1987, Greenpeace came to Geismar to march with local OCAW members, carrying "Bhopal on the Bayou" banners and other environmental warnings. This marked the beginning of what the OCAW called the "Toxic Watch Program," a campaign against BASF and other chemical companies to expose industrial spills, releases and poor environmental practices. One locked out BASF employee and his wife became rather famous for revealing those risks in their back-road and off-road "Toxic Tours. Darrell Stevens was a chemical technician who, along with his wife Ramona, regularly escorted reporters and environmental activists to the chemical plants, spill sites and Superfund waste sites strung between Baton Rouge and New Orleans, revealing data on toxins present, any history of accidental releases at each site and personal commentary on management practices and corporate attitudes in general. Darrell also began to use his training and skills to conduct daytime and nighttime surprise air tests in and around BASF and other chemical plants to document the exposure of unaware local residents to potentially harmful chemicals. As appreciative as local and national environmentalists and the media were for the information (I called on Stevens several times for stories, for example) in reality, Stevens had simply become the point man in the OCAW agenda to expose industry's poor environmental practices as part of the union contract negotiation strategy.

Later that year, local OCAW members flew to Germany to join the Green Party in their protests against BASF and other chemical companies. The contrast was almost incredible. The Green Party may have been one of the world's most radical environmental organizations, yet the usually ultraconservative, southern Louisiana hard hatters were rushing to join. But these were unusual times for local Louisiana workers, and the need for numbers and strength was paramount. Their strategy of revealing the chemical and environmental dangers posed by temporary employees opened a door to the local, national and international environmental movement.

When Greenpeace landed in Baton Rouge on its Mississippi River campaign during the summer and fall of 1989, it quickly re-contacted its local union and locked-out worker allies. OCAW workers joined Greenpeace and local enviros in the Great Toxic March from Baton Rouge to New Orleans. That unusual sight displayed just how far afield the unions had flung themselves to gain support and newspaper ink: young, hippie-like, Greenpeacers, complete with costumes, longish hair, shorts and sandals marched hand-in-hand with those tough looking, weathered, blue jeaned, hard hatters. If the year had been 1965, I would have half-expected to see the hard hatters start roughing up the hippies. Yet, along the chemical corridor of the Mississippi, I believe OCAW members were more than just tolerant, they were instrumental in planning and carrying out the infamous Greenpeace "environmental actions," where "action" teams chained themselves to industrial gates, plugged up waste discharges in the river and hung large flags from bridge superstructures. Greenpeace needed manpower, data, inside information on the plants and the lay of the land, and I believe the OCAW and the locked-out workers gave it to them. But why? First, any pressure on BASF specifically, or on the chemical industry in general, could be useful in improved working conditions, safety concerns and contract bargaining. After all, public awareness of industry faults and poor management practices would place a lot of pressure on industry to change, perhaps to even re-hire locked out workers. If continued pressure on industry were the only reason, though, OCAW would still be involved in environmental issues. I believe the OCAW had a second, more objectionable reason which was the real driving force behind the union's involvement: The union's participation in the environmental movement was merely a bargaining chip! The unions stood ready to call off its members and to stop the press releases, calls to news media, toxic tours, marches and everything else to ease public pressure. In short, the unions would drop out of the environmental movement if and only if BASF re-hired its workers. Well, guess what. BASF eventually entered into a new contract with the locked out workers, and suddenly the union participation in the environmental movement disappeared too. It was a huge hit to the local movement and a great loss of leadership, money, members, time and effort in one swift signing of a contract. Today union involvement is almost non-existent. The local OCAW environmental expert is gone and the phone calls to the news rooms from union members have ceased. Strike another blow to the movement. If I were a member of the local environmental groups who were still concerned about BASF pollution, I would feel sold out.

# Don't Mention Birds!
## Conservationists (and everyone else) go green

At first glance, conservation organizations would appear to be the most likely nucleus of the environmental movement. They had been working to preserve natural habitats for decades and had a very large following. Yet when these groups jumped into the movement, it was at a significant cost to their own organizations and eventually to the grassroots environmental movement itself. The Audubon Society, the Sierra Club, The Global Wildlife Fund, the Nature Conservancy and other conservation groups spotted early on the huge flocks of common folk migrating to the environmental cause and reached the same conclusion as the trade unions - there was membership and money in those crowds. Each organization was forced to act fast to recruit members and raise money for its own mission although, strictly speaking, few of them were truly involved in reducing pollution and cleaning up air and water - the underpinnings of environmentalism. The Audubon Society was known worldwide for its efforts to document and protect wildfowl and bird habitats. The Sierra Club had worked the political scene since its inception to preserve and protect unspoiled landscapes for the enjoyment of future generations. The Nature Conservancy's mission was the preservation of natural habitat, not so much through the political process as by raising donations, buying land and turning it over to suitable caretakers, and the Wildlife Fund was dedicated to preserving endangered animal species. To most moderates - the growing market for the conservationists - those missions sounded environmental, and the groups themselves seemed the perfect bedfellows for environmentalists. Certainly the idea of conservation was an environmental one and would fit right in with the grassroots environmental movement. But remember what the enviros were doing. They were clashing head on with polluters; picketing fence lines and corporate gates, holding toxic marches, joining with Greenpeace and others to delay or halt development, and they were filling legislative halls with stickered and placarded crowds. They were trying to shut down incinerators, close plants and clean-up toxic waste sites. None of those acts fell neatly into the missions of the conservation societies, which had not been that confrontational in the past.

After aligning themselves with the green movement, conservation groups found they were now lumped into the category of "radical environmentalists." Some groups, The Sierra Club for one, blended right in, others, like the Audubon Society, wound up spending a good deal of time protecting their moderate image so they could retain long-time members, engage in firm but friendly debates with corporate leaders, sponsor acceptable educational events to further their own goals and ask for donations from the corporations as well as conservative mom and pop environmentalists. Although it's got to be tough to rally big corporate support with one hand while holding a name-calling placard in the other, many rather high-brow organizations coveted the potential membership and contributions from the environmental crowd enough to fight for their share, choosing to repair any damage to their image after the fact. Their change in tactics, missions and messages ultimately harmed the organizations, because they increasingly appeared to be drifting in the political wind - not a great posture for fund raising. The rush to turn green also harmed the movement by blurring the goals and creating a tug of war for members and money rather than momentum to clean up the good earth.

Take the Audubon Society for example. Here was a world-wide, well known and supported organization, revered by bird lovers and naturalists across the globe. However, as its members aged and its image turned musty, membership rolls and income began to shrink. To swell its ranks, the group turned green to attract the enviro crowd. Audubon membership materials, literature and fund raisers suddenly began to appeal to the younger clean air, clean water crowd. The bird watching society suddenly appeared to be in the business of cleaning up toxic waste, and many members and observers wondered what happened. Even the president of The Audubon Society himself, Peter A.A. Berle, recognized that bird-watching and toxic clean-up were difficult messages to combine in a single breath. It was just too much of a stretch in logic and a difficult public relations message, so he found a cure - drop the birds! He employed a public relations firm which assessed society members' attitudes and reached the conclusion that birds were old fashioned. Responding quickly, Berle instructed his chapter officers to "refrain from overusing bird images." What? The Audubon Society down-playing birds? That's pretty indicative of how far afield the birders wandered to corral the grassroots environmental membership. Needless to say, long time birders and the long-time, mostly conservative Audubon members were not happy.

The Sierra Club, with its rather large library and catalog of conservation items for sale, entered the environmental movement in a variety of ways, including joining press conferences, responding to reporters' questions, and providing data on land use, shrinking wilderness acreage, logging, clear-cutting and such, all good conservation based strategies. About the time I began reporting for WBRZ, the New Orleans area chapter president, Darryl Malek-Wiley joined civil rights leader Pat Bryant and a few others at a press conference where they ceremoniously dumped a dead fish, presumably laced with one toxin or another, right on the lectern and announced they were going to push the state's health agency into doing something to clean up the lake it came from. I knew for sure the Sierra Club was entrenched in the movement when I saw an ad for one of their new books, published in 1995. Was the book about where to find and hike nature trails? Was it a guide to the national parks? Was it a map of bike and foot trails in Yellowstone? Was it about how to camp and hike with minimal impact to the woods you were in? Was it anything at all like the outdoor and exploration literature Sierra Club was famous for? Nope. The new book was a text in political and social science: *Deeper Shades of Green: the Rise of Blue Collar and Minority Environmentalism in America.* It was quite a stretch for the Sierra Club and was published, I believe, to appeal to the enviro crowd and the enviro buck. Suggested retail price: $30.

Even non-conservation groups like The League of Women Voters were turning green. At a time when women's issues and women's roles in government were big topics, you'd think the league would use all its resources to educate women on how to get involved, or how to work the political arena, or even how to run a political campaign. Instead, in the mid 80s the league used a good deal of time and money to publish a series of "Citizens Handbooks on Environmental Issues" entitled *The Garbage Primer*, *The Plastic Waste Primer* and *The Nuclear Waste Primer.* In fairness, I should state these books are well done compilations of the facts and issues of each topic. They are truly primers in that they compile the details, agencies, organizations, laws, facts and issues of each topic in a non-partisan, objective manner. However, the effort spent on these publications is just another example of the greening of organizations across the country in order to grab membership, donations and recognition from the grassroots environmental movement.

Greenpeace, although not a conservation group by origin, is a perfect example of an organization which did not hesitate, twice, to alter its calling in search of more members and more money. Greenpeace founders started their organization to stage protests against the proliferation of nuclear weapons and for their destruction. Their first "action teams" attempted to stop underwater nuclear testing by manning a school of small, fast, inflatable boats and wave-hopping into the test areas. Remember that colorful Greenpeace vessel *Rainbow Warrior* which would sail into the sites of underwater nuclear tests to attempt to halt the explosions? The French finally got so fed up with the interference of its tests, it deployed a navy underwater demolition team to sink the *Rainbow Warrior* while at harbor in New Zealand.

Later on, when whaling and hunting baby seals struck a nerve with the public, Greenpeace was there to stop that too. I remember the vivid footage of Greenpeace Zodiac teams buzzing around towering Japanese and Russian whaling vessels like so many flies, placing themselves between the vessels' harpoon cannons and the whales. Equally compelling footage from Nova Scotia showed Greenpeace members throwing themselves over baby seals in Canada to prevent fur traders from clubbing the seals to death. This was great TV stuff. Greenpeace was getting international news coverage everywhere it went, advocating the breaking of man-made law which they claimed was contradictory to the laws of nature and a clean earth. But was the organization an anti-nuclear group, an anti-animal killing group, or a group devoted to a new world order? Apparently, not even Greenpeace could decide, because it swooped into the new environmental movement, right along with everyone else.

Although the subject matter was different, the tactics were exactly the same: small bands of action teams would put themselves between polluters and the environment. We saw college students and young adults on TV, chaining themselves to the front gates of chemical plants in an attempt to prevent business as usual for a day, and even donning scuba suits to plug the underwater discharge pipe of a chemical plant. Greenpeacers also raced their Zodiacs outfitted with water pumps and hoses into a Louisiana bayou to suck up polluted water and spray it back onto the property of the culprit, an adjacent hazardous waste incinerator company called Marine Shale. Unlike earlier actions along the Mississippi River where unions and Greenpeacers worked cooperatively, in Amelia, Louisiana, hard-hatters clashed with the Greenpeacers. Marine Shale employees were waiting for the action teams along the bayou's edge armed with concrete reinforcing steel bars, commonly called rebar. The ensuing skirmish left one Greenpeace leader with a nasty head gash and a bloody face, and TV viewers across the state got close-up looks on the 5, 6 and 10 o'clock news. That very obvious, public fight with workers, who looked a whole lot like the ones in Baton Rouge Greenpeace was earlier trying to befriend, should have tipped off OCAW and local enviros to the real Greenpeace agenda. Greenpeace was not concerned about worker's rights, the image of Louisiana hard hatters or employment - it wanted to shut down industrial plants, period.

Because Greenpeace tactics, including forced entry into hometown environmental causes, illustrate my point so well, their actions deserve a close look. In the summer of 1988, Greenpeace launched its Great Mississippi River Campaign and locked-out union members were there to help. Greenpeace had bought an old tugboat, armed it with water sampling equipment, a computer, and those famous inflatable, high speed Zodiacs and sent it down the length of the river, stopping at various chemical plants along the way to sample the water. A radio filled Greenpeace bus, used as a mobile headquarters and barracks, followed along on the river bank. At every stop, Greenpeace teams cleverly manipulated the TV crews and press reporters to get dramatic coverage. In addition to those prime time moments of pollution-stopping actions, Greenpeace also gained some ink and TV time by joining local grassroots environmental organizations which were holding rallies, marches, press conferences and fund raisers. Even though local environmental leaders and grassroots members were already quite leery of Greenpeace tactics, they succumbed to the pressure, allowing transient Greenpeace teams to set the agenda, build awareness, then grab members and donations. As I observed it, this was the beginning of the split in the local environmental movement and reflected what was happening across the nation. Local organizations became divided. Some were pro-Greenpeace, almost militant, wanting the conflict and controversy, while others were decidedly against, seeking more moderate tactics instead.

Greenpeace's actions in Baton Rouge were a perfect example of their expert media manipulation. In the newsroom, we knew Greenpeace was making its way down river and would stop for a while in Baton Rouge; Greenpeace faxes and mailers had told me so. When a Greenpeace advance team arrived, it quickly rounded up news reporters' numbers from our local enviro leaders and started calling to fill us in on what they were doing. The pattern was this: Greenpeace members would call to say they were planning something big for Baton Rouge; a few days later, they would call again to say something like, "Be at so and so place in an hour. Something's going to happen there." Sure enough, at the appointed time, Greenpeace teams would appear and start their action, letting reporters, photographers and TV cameras get prime footage of the well-choreographed event.

In one case, the call came to be at a downtown office building called One American Place in an hour. Since we all knew that was the headquarters of the Louisiana Chemical Association, we all rushed down to see the excitement. Sure enough, Greenpeace then rolled up with a rented crane to hang an anti-chemical company banner on the building. We got good pictures, including the arrest of the Greenpeace team and the chemical association personnel looking a little bent out of shape, all of which was planned perfectly by Greenpeace. However, when the chemical association personnel didn't respond with enough anger to make themselves look foolish on TV, Greenpeace provoked them more. During a TV interview I was conducting with the spokesperson for the chemical association, the Greenpeace point man sidled up beside my interviewee just to annoy and harass him during the interview!

Greenpeace skillfully played with those of us in the media partly because we could never locate them between actions. They took up residence on the outskirts of the city, or so I'm told, for weeks, but the only contact I had with them was during their actions or during the phone calls or visits they made to let me know about their next action. Granted, I did not have the luxury of time to look very hard. Even though I was on the environmental beat, I was not allowed to simply hunt down Greenpeace. Maybe I could have found them if I looked harder, but there were other stories. Greenpeace knew this too and played on that weakness. They behaved exactly the opposite of the local grassroots organizations; they were secretive, confrontational and not available to the media except on their terms. We could only cover them when they wanted it. And if we broke their rules, we were subject to being cut off from their information flow.

I learned the price of breaking Greenpeace rules during the biggest action of their 1988 Baton Rouge stop. They had planned to block the underwater Mississippi River discharge pipe of Georgia Gulf, a chemical plant near Plaquemine, Louisiana. My assignment editor and I, along with my news director, knew Greenpeace was fond of using decoys and staging events to throw off the local authorities to gain time for their real action. So, we worked out a code system where I would make a single, very quick cellular call from the site of the action to tell my assignment editor which chemical plant was the real thing. We took time to work up that quick code to ensure balanced coverage to our viewers in two ways. First, we were concerned that Greenpeace might not grant us, or any other reporters, access to company spokespersons at the site. We just didn't want to be stuck on Greenpeace's boat with only a single viewpoint, that of Greenpeace. Secondly, we wanted another cameraman shooting from the company's viewpoint on land. We had learned our lesson; although we were subject to Greenpeace manipulations in some actions, in this one we would be in total control of the coverage. So, we put a helicopter and second videographer on standby to respond to my coded message if I could send it.

An afternoon later that week, I got a call from Greenpeace to be at the Plaquemine ferry landing early the next morning because, "something interesting will happen." There it was. The big action we had been waiting for. Videographer Bob Gray and I drove to Plaquemine the next morning to join a crowd already gathered there, including a few local sheriff's deputies. Before long, the Greenpeace boat, Beluga, pulled up to the landing. Reporters were invited on the boat for a ride, the deputies were not. Once aboard, and well away from the dock or any other means of returning to the newsroom, Greenpeace leaders called a meeting on the bow. That's when we learned that a diversion team with a large, fake plug was simulating an action several miles further downstream from us near BASF (as I recall), and the real action team was in the water just ahead trying to plug up the underwater discharge pipe of Georgia Gulf. The head Greenpeacer that day, John Liebman, also laid down the law for us: there would be no transmissions of any kind from the boat from that moment on.

Great. Greenpeace had just told us the location of the real action and the diversion - a great story for anyone with two cameras to cover them both. But, Greenpeace had also just told me how to do (or not do) my job. Here's how I looked at the situation. Greenpeace had invited me on the boat to cover a story. They laid down no ground rules for boarding and asked for no agreements from me as a condition to board. I showed up as a reporter and they let me on, presumably with all reporter's tools allowed, including a cell phone. Greenpeace conducted no bag searches, asked no questions about gear and mentioned no restrictions of any kind prior to boarding. Then, in the middle of the Mississippi River, miles from the newsroom, Greenpeace leaders announced their rule against radio or phone transmission of any kind and proceeded with their briefing of the current action. That's when we all learned the details of two news events: the diversion and the real action. At that point I had enough information to practically write the story as well as get help to cover the diversion. To me, calling the helicopter to cover the diversion was critical for two reasons: first, it may have been an illegal interference with law enforcement, second the Beluga would not be going there. So, I did what any gentleman reporter would do. I snuck away from the briefing, ducked as low as I could behind some large tugboat contraption and made the coded call. I might have gotten away with it cleanly, but cellular service wasn't very good from the middle of that stretch of the river, and I couldn't get through directly to the newsroom. I had to call my wife at home, and to her surprise, hurriedly asked her to skip any questions and just relay the simple message: "Diversion at BASF, real thing at Georgia Gulf."

Just as I was finishing the call to my wife, I looked up to meet the eyes of a very angry looking Greenpeacer. I hung up and mumbled something about my darned cellular phone and how hard it was to hear my wife clearly. He looked as though he would actually throw me off the Beluga right into the middle of the Mississippi River, before I reported his story. He got even madder when the cavalry arrived - my television station's helicopter. It buzzed us on its way to the diversion with videographer Martin Deroussel leaning out, rolling tape. When it returned, we were anchored just offshore from Georgia Gulf as more kindly Greenpeace activists in inflatable Zodiacs began ferrying photographers and videographers right up to the colorful action team for great close-ups. Wearing bright red dry suits and standing in chest deep water, some action team members had erected a hand-painted sign with skull and cross-bones declaring the area a toxic site and were granting interviews. Other action team members were granting interviews from their own inflatable boats. It was kind of nutty, really. Videographers and reporters with microphones leaned out of small inflatable boats to get pictures and sound from characters in other boats, while just a few feet away people dressed in bright red held up that silly banner while still others fought against the flow of the discharge pipe to clamp a very home-made looking hatch on it. Great stuff for TV, and for a while there it looked like every reporter was going to be given a ride but me. Imagine that. When I mentioned this to the boat crew and voiced some concerns about how Greenpeace wasn't letting me do my job, I was told to get in an inflatable for a closer look, but I would have to come straight back to the main boat. I could not get off on the river bank.

Me: "Why? I need an interview with the Georgia Gulf people up there on their dock."

Greenpeace: "If we drop you on the bank, we won't be able to come get you,"

Me: "So?"

Greenpeace: "You'll have to find your own way back to your car."

Me: "Fine. If that's how you want it, let me go.

And that's what happened. Bob and I got in the inflatable, got some good video and sound. Then I climbed ashore, thanked our Greenpeace taxi driver and left Bob to continue rolling on Greenpeace while I met Martin onshore. We got interviews with the Georgia Gulf employees on the dock, dashed for the helicopter and flew home. The Beluga took an hour getting back. We were back in 5 minutes. See you later Greenpeace! I'm just sorry Bob Gray had to ride back on the boat. The good thing was he continued to roll, getting good video of the diversion team returning with their fake plug.

    I did two stories that day. The first was on the action itself. Greenpeace, despite its protests to me about making the call, actually got better coverage from us than from the other stations because we had two cameras. With one on the riverbank and one on the Zodiac, we captured a shouting match between landlocked sheriff's deputies and the waterborne action team. Our 6 o'clock news viewers saw the Greenpeace leader calling loudly to the sheriff onshore, accusing Georgia Gulf of trying to blast the divers by cranking up the pressure on the discharge pipe. He claimed the act was dangerous and intentional. The Sheriff, equally loudly, told Greenpeace if they didn't stop, he'd take all their boats away from them. In the news business, that's called a scoop. We were the first to report a potentially dangerous act on the part of Georgia Gulf. We put Greenpeace leader Brian Hunt on the air shouting, "It's just like putting a gun to their heads and pulling the trigger, officer." We then ran to the dock and confronted Georgia Gulf employees with the Greenpeace accusation. They said that even if they wanted to goose the divers, they had no means to turn up the pressure.

    The second story I reported, at 10 o'clock, was about the Greenpeace media tactics and manipulation, including their order not to make a call from the boat. I guess that story didn't go over very well, because despite all the great coverage our helicopter and second videographer provided, Greenpeace was highly agitated over my actions. The next day I had to explain to my boss, news director John Spain, why he just received a complaint from Greenpeace concerning my behavior and their decision to leave me off their call list for the rest of their stay in Baton Rouge.

Greenpeace claimed I jeopardized their mission and endangered their action team. Fortunately, John agreed with me. He left me on the beat. The reality was Greenpeace needed our TV coverage just as badly as ever and couldn't blacklist me for too long. But every encounter I had with Greenpeace after the Georgia Gulf incident had a distinct chill to it.

Here is, I think, the strange thing. As manipulative as Greenpeace was, ours was the only station in town to buck their system and report exactly how they operated. That's how clever and compelling the Greenpeace actions were. They drew us reporters in, kept us in and got great coverage. Wow! What drama! What action! TV stations, including my own, and newspapers couldn't ignore it. This was one colorful set of characters. Their stuff was exciting, complete with video, pictures, confrontation and dangerous activities, all newsworthy stuff. Frankly though, I was surprised other reporters didn't cover the story on their own rules, rather than those laid down by Greenpeace.

The resulting news coverage was able to do one thing very well, though; it made Greenpeace a household word in Louisiana and helped build some support for them among viewers. That support was apparent during much of Greenpeace's Mississippi River campaign, where impressive crowds gathered along the way. The Greenpeacers, somewhat hippie-like creatures, driving a large bus and piloting the water-sampling boat Beluga, made their way from crowd to crowd down the Mighty Mississippi, chaining themselves to chemical company gates (illegally, I might add), zipping along the river in Zodiacs in an attempt to plug up industrial water discharges (also illegal), and spending days (illegally) in the superstructure of a huge bridge over the Mississippi River in Baton Rouge trying to hang an anti-chemical company banner. It was fascinating to watch the rather conservative crowd of granddads, grandmoms, moms and dads, who were generally devoted to law and order,

## Goodbye Green                                    Duncan

cheer on the illegal acts, and join many of the more traditional ones like the marches, rallies and public hearings. Greenpeace also joined hands, literally, with local grassroots organizations and unions to stage the Great Toxic March from Baton Rouge to New Orleans. Hundreds of people, many in costumes, started at the head of "Cancer Alley," near the little town of Alsen, which lay just across the street from Rollins' hazardous waste incinerator, and spent the next week making their way down River Road to New Orleans. There was even a touch of symbolism to the march. River Road was the same route made famous by the anti-establishment motorcycle trek made by Peter Fonda, Dennis Hopper and Jack Nicholson in the movie Easy Rider.

The Toxic Marchers stopped now and then at the entrances to several chemical companies to shout and cajole whoever might be in earshot at the time to clean up and stop polluting. Curiously, the moderates and conservatives in the crowd had been asking to see the inside of these places for years by requesting meetings, tours and the like. But during the march, surrounded by strange, costumed characters, hard core Greenpeace action teams and radical activists, moderates seemed unusually content to stand quietly with their signs. Although more than one corporate executive stepped outside to greet the Toxic March with the exact invitation many moderates had wanted to hear for years - to come inside and talk - the marchers refused. Such was the power and persuasiveness of the new, radical leaders of the movement who preferred shouting to talking. They were beginning to dictate tactics, and grassroots moderates were following. The effort eventually rang rather hollow, however, because the Greenpeace teams would stir up a lot of emotion at each stop and then leave. Those moderate grass roots members left behind in the swelled, anticipating crowd soon found themselves asking "what next?"

Then, Greenpeace unknowingly took that mild confusion and whipped it right into anger with a single action - the last of its Mississippi River campaign. One night, a small Greenpeace team climbed into the superstructure of the largest Mississippi River bridge in Baton Rouge. This particular bridge bore the traffic of Interstate 10, the major artery connecting the east and west banks of the city, not to mention the east and west coasts of the United States. Greenpeace wanted to turn the crush of automobile traffic into one huge audience, so action team members planned to bivouac mountain climbing style for a few days and at dawn on the first morning, unfurl a banner which read "Cancer Alley Brought to you by _____ ." They then planned to fill in the blank each day with the name of a different local company. The problem started that first night when the temperature plummeted and the wind howled. By dawn's first light, the Greenpeace team's banner had become twisted, snarled and almost unusable. Then the local police showed up. While the Greenpeace team struggled in vain to pull the banner in, untangle it and re-release it, the local police carried out the same procedure they have always followed when someone climbs the bridge, either as a stunt or a potential suicide. Officers blocked off the lane beneath the people on the bridge, shifted traffic to other lanes and tried to talk the climbers down.

Thus began a two-day stand-off which was widely covered by the news. Sure, Greenpeace reached the population of Baton Rouge and surrounding communities through dramatic TV and newspaper coverage. Hundreds of thousands of viewers and readers saw that large, flapping, slightly crumpled banner which read "Cancer Alley brought to you by DOW ." (The Greenpeace team had managed to fill in the single name, but could not reel in the wind torn, twisted banner to change it again.) More importantly, though, Greenpeace had just broadcast to the entire population of Baton Rouge just who the heck was causing the miles-long traffic jam leading to the bridge. Let me tell you, the crowd turned ugly. As a reporter on the scene, I saw that swing in mood up close and personal. What once seemed to be a town in favor of Greenpeace became little more than a long line of hot, mad drivers and passengers snarling through open windows everything from "Get out of town," to "Drag 'em down out of there," to "Just shoot 'em." Then, from the midst of the jammed cars, frayed nerves and Greenpeace action, an event occurred which poured salt directly into the public wound: a motorist startled by all the activity right in front of him, slammed into the rear of a parked police car, put there to close one lane and direct traffic into an open lane. The next thing Baton Rouge saw and heard on TV was the highly respected, widely recognized and very angry Sergeant Carl Thompson adding up the monetary loss to the taxpayers of Baton Rouge in both manpower and vehicles. It was probably good that Greenpeace ended its local campaign right after the team climbed down from the bridge and police hauled them off. Any more publicity like that and any remaining Greenpeace supporters may have turned in their membership cards and asked for their money back. That single event caused Greenpeace to lose substantial support from the Baton Rouge moderates and created an important lesson for all wannabe environmental groups. Pay attention, I call this the *Greenpeace Rule:* The quickest way to turn supporters into an angry mob of opponents is to make them late to work by creating one huge traffic jam during morning rush hour.

Greenpeace, like the other national pseudo-environmental groups, had jumped on the green bandwagon, grabbed membership and dues while it could make local headlines and then moved on. At precisely the same time, the Audubon Society, The Global Wildlife Fund, The Sierra Club, The Nature Conservancy and all of the other conservation organizations were appealing to the green crowd as well. This divided the money, the time, and more importantly, the attention of grassroots environmentalists and contributed to the loss of direction of a good movement.

## Goodbye Green — Duncan

Now here's the latest. Greenpeace, on September 15, 1997, announced its closure of all but one U.S. field office and a staff cut of 85 percent. The result was 300 of about 400 door-to-door canvassers were put out of work. Canvassers, by the way, were those young people who knocked on our doors, aroused our anger at the establishment and asked for donations to stop whaling, nuclear bombs and pollution. Greenpeace regarded them as true front line foot soldiers in a war - an environmental war - against the establishment. Greenpeace officials said the canvassing cost more money than it brought in, and the organization faced a $2.7 million deficit. Stop and think about that for a moment. Greenpeace had become, probably, the world's single most recognizable environmental organization, yet soon after, the group no longer had enough money to stay open.

Door-to-door canvassing is the backbone of a grassroots campaign, but Greenpeace canvassers were having trouble getting everyday Americans to chip in. That was a sure indicator the extremist group had completely lost the moderate, grassroots membership which made the environmental movement so strong. The reality was people had been dropping out of the movement and abandoning Greenpeace for years. In 1997 its budget and its number of donors had dwindled to a third of the level the group enjoyed in 1991. Times had turned bad for the organization. The chief spokesperson admitted that during the environmental movement's golden years (the late 80s) money and membership had rushed in so fast, and the organization had built up so quickly, that Greenpeace leaders were unable to keep up with it all. They had no plan for the future and certainly no way of maintaining the huge number of members who heard the knocks, answered the doors and responded with donations in the 80s. Leadership didn't recognize the simple fact that the majority of donors and new members were moderates, blindly rushing into the organization in an effort to do something, anything, to help the environment.

Greenpeace accepted the money and people, but never catered to a moderate membership. Hanging from bridges, tying up traffic, getting arrested, provoking shouting matches and diving into murky waters to block off a corporation's legal discharge pipe were not the sort of activities with which polite folks wanted to associate. The polite moderates left and took their money. Greenpeace leaders, still basking in their newfound mass popularity and a swollen bankroll, had no plan for hanging onto its moderate members. One day Greenpeace simply woke up, saw the unpaid bills and closed shop. Don't believe me? Listen to the people on the inside.

Bill Keller, acting executive director of the U.S. office of Greenpeace admitted in national news stories, "Our membership has declined, our supportership has declined and our average budget has declined."

Barbara Dudley is the former executive director of the U.S. office of Greenpeace. In a National Public Radio news story about the closings she said, "We were willing to just engage in this thing that we now look back and scoff at called 'churn and burn,' which is to just churn members through because they were coming in such enormous numbers. And you have to remember that this is an organization of anarchist activists, and when millions of dollars started rolling in, nobody thought of the future."

Anyone still arguing with me that the enviro extremists' agenda veered widely from the moderate and conservative membership of the grassroots movement?

Ironically, some extremist observers felt Greenpeace lost its membership in an attempt to become more mainstream. They say the organization spent more money writing legislation and lobbying than on protesting. According to Earth First! founder Dave Foreman, Greenpeace leadership put the viability of the organization above its cause. He called it a "growing breech between grassroots volunteer activists and professionals." (Confessions of an Eco-Warrior, Harmony Books, 1991). That may be a valid point supported by the fact that the only U.S. Greenpeace office left open is its Washington D.C. office. However, I believe the mainstreaming of Greenpeace - if that is a proper characterization - was a private battle which divided the establishment oriented front office from the radical mid-level leaders. Greenpeace's continued public extremism is what sent its moderate members overboard.

In an analytical column for the Knight Ridder Newspapers group, Christopher Boerner and Jennifer Chilton-Kallery made the point that as the Sierra Club, National Audubon Society, the Wilderness Society and others swelled in membership and diversified in goals, the front offices tended to lose touch with members. Some local chapters even revolted. And the newly diversified organizations tended to spread into common territory, offer the same services and thus compete for the same membership dollars and donations. This led to the general fall off in support in the early 90s, so these organizations started to try to differentiate themselves with exaggeration and hyperbole. According to Boerner and Chilton-Kallery, "the more striking and exaggerated the predicted calamity, the greater the likelihood reporters and, in turn, politicians and potential contributors, will take note."

It is interesting to note that even as Greenpeace was closing its U.S. offices and bemoaning the loss of its grassroots environmental giving, the organization was protesting yet another cause, the genetic engineering of food. This is truly an organization that shifts with public sentiment to stay in the news and in the pockets of Americans. What's more, I even question Greenpeace's own philosophical underpinnings. It may be an organization of more show than substance. As I was touring the Beluga in Baton Rouge, right there next to the lap-top in the on-board laboratory I spotted

3 1/2 inch computer disks (paid for with donated money, presumably) manufactured by none other than BASF. Apparently - and ironically - not even Greenpeace could function without the benefits of modern plastics.

Other organizations seemed to have lost their way completely. One example is the Streisand Center for Conservancy Studies in Malibu, California. Barbra Streisand (yes, the entertainer) created the center and later donated it to the Santa Monica Mountains Conservancy in 1993. Nearby residents claim that ever since, the site has been the center for noisy commercial events like weddings, tours and parties and filed suit to stop the activity. The homeowners group president was quoted in local and national newspapers as saying, "To date, we have no record of a single study or significant work concerning nature or the environment."

There are some books which reveal the inner workings of some rather extreme groups and the reaction their extremism caused. *In Act Now, Apologize Later* (Cliff Street Books, 1997), Adam Werbach, the youngest Sierra Club president, reveals how his organization and the environmental movement have become merely a youthful, extremist cause. *Trashing the Economy* by Ron Arnold and Alan Gottlieb (Free Enterprise Press, 1994) reveals how so-called environmental organizations like Greenpeace, the Nature Conservancy, the Audubon Society and others really didn't live up to their claims or to their stated goals. You might also check out *The Green Revolution* 1962-1992 by Kirkpatrick Sale ( Hill and Wang, 1993) and *Green Backlash* by Jacqueline Switzer (Lynne Rienner Publishers, 1997). Another Greenpeace watcher is Mark Dowie, author of *Losing Ground - American environmentalism at the close of the twentieth century*, (MIT Press, 1995) in which he states his belief that the environmental movement became ineffective when corporate America got involved.

# Green Justice

## The Civil Rights Movement Crowds Out The Environmental Movement

Early on in the environmental movement, scholars predicted that the black community would never be a loud or strong voice for clean air and water because for most blacks the issues of jobs, equality and social security were much more pressing. J.A. Swan wrote in 1972 that "it is doubtful that air pollution will ever become an extremely high priority item with the very poor. Only limited success may be possible in efforts to convince urban blacks that pollution is worth fighting" (in *Environment and the Social Sciences: Perspectives and Applications*, American Psychological Association). A few years later John M. Ostheimer and Leonard G. Ritt echoed those sentiments saying low-income minorities have too many other things to worry about (*Environment, Energy and Black American*s, Sage Research Papers in the Social Sciences, 1976.)

Those scholars were wrong in their predictions. Blacks and minorities wanted clean air and water like everyone else and readily joined the environmental movement alongside whites, Hispanics and other minorities. They were blue collar, white collar, private sector and public employees who literally marched down streets hand in hand in a movement which seemed to be truly cooperative and non-racial.

It was no coincidence, however, that the successful entry of minorities into the environmental movement occurred at the same time some black leaders were bemoaning the decline of the civil rights movement. You may recall vocal efforts from black leaders throughout the 80s attempting to pump up the civil rights movement, saying it had lost momentum due to large numbers of blacks dropping out. Apparently the dropouts had either achieved what they wanted in job status, political strength and social needs or they simply saw the movement as no longer relevant. After all, blacks had made huge gains in this country. They and other minorities were successful at all levels of government, business, education and entrepreneurism. Whatever the reason, civil rights had become a weak issue to black conservatives and moderates. The result, I believe, was that the problem for leaders of the NAACP, the ACLU and local minority interest organizations became how to stimulate more interest in lingering civil rights issues and re-invigorate the civil rights movement.

Ben Chavis almost single-handedly solved that problem when he coined the term "environmental racism" while serving as head of the United Church of Christ's Commission on Racial Justice. In 1982, Chavis joined citizen protests against a North Carolina landfill. He claimed the landfill company chose the site racially, meaning the company intended to place the landfill near a small, black community specifically because its citizens did not have the political and economic clout to stop it. Chavis' shouts of racism concerning the site selection of a potential environmental hazard eventually gave birth to an entirely new mini-movement against environmental racism, and in 1987 the Commission for Racial Justice issued a report on the subject entitled *Toxic Wastes and Race in the United States.* It's lead author was Ben Chavis himself, and he used the term environmental racism to describe the report's findings and carried the issue with him when he became the head of the NAACP in 1993. Participants and bystanders alike saw this as a major step in the environmental justice arena. Tom Soto, director of Los Angeles' 1990 Earth Day and president of the Coalition for Clean Air, spoke of the Chavis/Commission report as "a seminal report delineating these truths" about racial injustice in minority communities and said Chavis' appointment to the NAACP "promises to further strengthen the (environmental justice) movement.

During the decade following the Chavis/Commission report, the term " environmental racism," and the mini-movement behind it, began to evolve as black leaders searched for the perfect issue to advance their civil rights agenda and to regain support of moderate blacks and more whites. However, environmental racism, as a potentially unifying issue, proved to be too narrow, too difficult to prove and too inflammatory to gain wide support of black moderates and non-minorities. So the search began for a broader issue and a softer term. The first evolutionary step occurred about 1990 with the term "environmental equity term which apparently meant that both economic benefits and environmental risks of polluting industries should be shared equally among the races. For example, this position held that it was environmentally iniquitous for a large chemical plant to hire mainly whites, while sending pollution into nearby black communities. The beauty of environmental equity as the new issue was that blacks did not have to prove intent to harm, meaning they did not have to prove actual racism. Blacks merely had to claim they breathed bad air while whites got the jobs. That shift of argument from environmental racism to environmental equity worked. More whites and environmental leaders began to line up behind black leaders in their fight for environmental equity for poor black communities near industrial sites.

But, like the earlier term of environmental racism, the new term, equity, eventually became a tough sell as well. Apparently, the ideal of equal environmental rights (environmental equity) was an issue, or term, which was too narrow in the overall civil rights agenda, which included all rights not just the rights to clean air. Also, to make any advance toward environmental equity, blacks still had the very difficult task of proving environmental inequality. So the next evolution occurred; one more step away from clean air and water, and one more step toward jobs, better pay, and improved opportunities for blacks. The new, potentially unifying phrase was "environmental justice," the notion that industries which had harmed (that is, polluted) poor black communities in the past now faced the responsibility of correcting those harms. The idea of environmental justice also held that industry had an active responsibility to ensure that life in and around chemical plants was fair and just, regardless of industry's original roll in creating any injustices. Did you catch that? Civil rights leaders were using the environmental issue to put pressure on local companies to right wrongs, create equality, improve life and bring fairness and justice to its surrounding communities, even though the companies may not have actually caused the wrongs, inequalities or injustices. In short, environmental justice as a battle cry was simply a strategy to rally blacks, other minorities and whites who lived near chemical plants to a comprehensive civil rights agenda. That agenda was to press industry and government to pay for correcting injustices in poor, minority communities near industrial plant sites. Sure that meant clean up the air and water, but it also meant pay huge taxes, build schools and roads, give money to the community, promote blacks into management, work toward electing minority representation, provide lots of materials and money to local schools, work to improve opportunities for minorities or buy everyone's home and relocate them - a whole gamut of civil rights and social issues.

Thus environmental justice became an effective strategy for blacks because it emerged as the central minority issue shortly after the grassroots environmental movement had achieved huge social and political gains; any agenda coupled with the word "environmental" was bound to be a sure success. Sure enough, the civil rights crowd so successfully captured the attention of President Clinton in 1994 that he created the Working Group for Environmental Justice within the Environmental Protection Agency. That working group was to oversee a new presidential mandate for every single federal agency to seek environmental justice. Here is his order:

> "To the greatest extent practicable and permitted by law, and consistent with the principles set forth in the report on the National Performance Review, each federal agency shall make achieving environmental justice part of its mission by identifying and addressing, as appropriate, disproportionately high and adverse human health or environmental effects of its programs, policies, and activities on minority populations and low-income populations..."
> Executive Order 12898
> - President Clinton

It's important to note the working group on environmental justice didn't report to the head of the EPA, but to the President himself. The sheer breadth of the order was greater than anyone anticipated.

Success! The leaders of the black civil rights movement had captured the environmental agenda, along with its grassroots members, support and political impact. Black leaders were then able to slowly but surely squeeze environment out of the agenda and replace it with a more traditional civil rights position. The early scholars, who predicted no involvement by the minority community, failed to predict the ability of civil rights leaders to turn clean air into a racial issue. Once poor minorities were told clean air and clean water was easier to obtain if you were white, the ranks of grass roots environmentalism swelled with minorities.

Unfortunately for the movement as a whole, the strategy ultimately backfired. The use of the terms environmental equity and justice became so far removed from actual environmental issues of clean air and water that many long-time environmentalists steered clear of the terms, sticking instead with their primary goals of clean air, land and water. One local and highly recognizable environmental leader told me, "I don't touch that issue (environmental equity) because it has become so distorted." That was Marylee Orr, head of the Louisiana Environmental Action Network, who said she'd rather concentrate on better defined subjects like clean air and water. Also, a local labor leader once told me, "Everyone defines environmental equity differently." Those comments were actually rather mild compared to those of Janice Dickerson, a woman in Louisiana who was intimately involved in the environmental justice movement.

Dickerson was the coordinator of the state's Community - Industry Relations Group (formerly the Environmental Justice Group), an office within the state's Department of Environmental Quality. Dickerson's job was to be a resource to communities in their pursuit of environmental justice. She accused predominately white environmental groups, the EPA and even black academicians of using environmental justice as a "special interest tool to generate funds for mainstream organizations," and said the issue "is being pimped for their own use." Her concern is that poor, black residents facing off against a large polluter didn't need their hands held by outside groups forcing their way into the community fight to further their own interests. She explained, "The manner in which the term environmental justice is currently being used by traditional white, large environmental groups and the EPA is that it is forced upon the people who count," meaning those who actually lived near chemical plants. Her statement to me was that the only people who should have a voice about environmental equity or justice are citizens who actually live near a large industrial site and who are affected by it, and that outsiders should enter only when invited. "Environmental justice is the prostitute anyone uses for their own cause," she told me, "and the EPA is allowing white environmental groups to move into communities and control them. The federal government allowed people affected to be put out of the process."

Her comments were significant and somewhat prophetic, because a few years later local black and minority citizens began to break ranks with the national environmental justice leaders, the U.S. Justice Department and the EPA over the location of a new chemical plant in rural St. James Parish, Louisiana. The locals wanted the jobs, Greenpeace, the Department of Justice, the EPA and the Tulane Law Clinic wanted to stop development, making the company a test case for President Clinton's new environmental equity directive.

So, here's the picture. For years in this country, civil rights leaders had been waging war against big business by claiming blacks had been left out of corporate leadership, creating a form of disenfranchisement. Then the environmentalists began targeting large chemical companies as polluters. What a fit! The environmentalists had opened fire on the exact companies the civil rights groups were battling, so the civil rights crowd rushed in, like the cavalry, to take up the environmental flag. But while waging a highly confrontational "us against them" strategy with large chemical companies, black leaders and civil rights organizations wrested so many environmental issues out of the hands of grassroots locals that many local black residents and other citizens with more basic needs - like jobs, good schools and adequate housing - felt abandoned. Ironically, in their zeal to fight companies on an environmental equity front, some civil rights leaders - in the eyes of locals - were actually fighting jobs and economic growth.

Look at the recent, dramatic split between residents of St. James Parish, Louisiana and "outsiders" over a large company called Shintech, Inc., which wanted to build a polyvinyl chloride plant in Convent, Louisiana. Environmentally speaking, Shintech clearly fit the potential big polluter profile, so when radical enviros sounded the alert, the cavalry raced to Convent. The EPA, Greenpeace, the NAACP, Tulane University's Environmental Law Clinic (a predominantly white university, but very pro-civil rights and pro-environmental equity), the Louisiana Environmental Action Network and the Rev. Jesse Jackson all rushed in to stop Shintech and pick up a few headlines. Sure, some local grassroots environmentalists were there, but so were the out-of-town big guns. The tactics were predictable: claim environmental racism; shout for environmental justice and start challenging everything from permits, to the fairness of public proceedings, to the motive of the company and motive of residents in favor of the plant; bring in the lawyers (thanks to Tulane) to start tangling the whole process up; then call the media to raise cane.

By the summer of 1997, I suppose, enough finally became enough. Louisiana's governor decided to take a look for himself. Governor Murphy "Mike" Foster had been reading the papers and watching the news like everyone else. He heard the opposition clearly, but also clearly noticed most of the hurrah he read about and heard seemed to be coming from folks who didn't exactly live in Convent. So he doffed his tie, threw on some jeans and drove on down unannounced, just to meet the folks in town and hear their side of the story - without the normal gubernatorial hoopla and entourage. Turns out ol' Mike had a hard time finding anyone opposed to the Shintech plant. Area residents wanted the jobs, the taxes, the economic growth and the opportunities. They gave the guv an ear full, saying they were tired of outsiders trying to fight off industrial development by swinging the EPA's new environmental equity club. If the closing down of Greenpeace hadn't sounded the death knell of an environmental movement which had historically appeared to speak with one voice, then the chain of events in the weeks following the governor's Convent trip did. Governor Foster returned to Baton Rouge, announced what he learned - that town residents actually seemed to want the plant - and watched the fireworks begin.

"Foul!" cried the radicals, "He's wrong. Nobody down here wants that stinky 'ol plant."

Some residents agreed, I'm sure, but in September, 1997, just about the time Greenpeace was closing its offices, leaders of the Louisiana chapter of the NAACP announced during a press conference their concern that white people were using the black issue of environmental justice to head off the opening of Shintech, Inc. Ernest Johnson, head of the Louisiana NAACP, said that rather than working so hard to keep companies from opening in town, "our white friends in the environmental movement" should join blacks to fight other injustices such as police brutality and job and housing discrimination.

It's important to note here the leadership of the state's major environmental organizations have always been predominantly white and middle class. They were regularly joined in their fights by members of the Tulane University Environmental Law Clinic, law students who were also predominantly white and distinctly middle and upper class. Tulane's law clinic, like those at many law school clinics, was established to provide legal help to the poor, presumably in the usual matters of housing rights, job rights, personal injury and the like. But the Environmental Law Clinic opened specifically to wage big time legal battles against national and international corporations on behalf of entire communities - not just individuals - under the guise of environmental justice. Clinic members joined with community groups and outside environmentalists to fight

Shintech's plans for a new plant. They argued the company's management knew it would face strong opposition politically and legally if it tried to place its plant near a white, well-to-do community, so the company purposely picked a poor black town as its site. The problem that Governor Foster had pointed out, of course, was the local, predominantly black citizenry of St. James Parish actually wanted the plant and its jobs. Yet environmental organizations, aided by the Tulane Law Clinic and the U.S. Environmental Protection Agency, ignored the group of citizens who wanted the plant, and continued to try to stop the development under new environmental justice rules of the EPA. The schism between the agendas of the environmental justice flag wavers and those they supposedly tried to protect widened to a unbridgeable canyon at the NAACP press conference. The NAACP, which truly represented poor blacks, was saying "enough."

Wow! Let me tell you how significant that little break-up was. For years, the black community had provided extremely active and vital participants in the grass roots environmental movement. With its decades of experience from its own grassroots civil rights movement, they immediately brought the substantial weight of local organizations, churches and leadership to bear on environmental concerns. For Louisiana's black leadership to break ranks in Convent was a major blow. A greater blow, in my opinion, came shortly after when the Black Caucus of the Louisiana Legislature began calling for fewer regulations from the Environmental Protection Agency, which required the state of Louisiana to consider "environmental equity" issues before issuing permits to companies like Shintech. The black caucus believed too many EPA rules and regulations burdened small business development and prevented industry growth and therefore prevented jobs. Apparently, the Black Caucus and the local NAACP were saying the same thing: the issue of environmental justice (or environmental racism) was of less concern to the black community than the betterment of lifestyle through new jobs. Both groups stated or implied that the environmental racism issue was being used by white, well-to-do environmentalists who didn't live in St. James Parish in order to further their own interests.

Soon after that, the head of Louisiana's NAACP announced that he was actually for the Shintech plant and the jobs it would provide. In fact, he helped broker a deal which, in my mind, was exactly what small town residents living in the shadows of large industry had been clamoring for. He worked with a group of business people to craft a proposal they used to seek a business development loan. The proposal was attractive enough that the state of Louisiana pledged an additional $2.5 million in matching funds. The group wanted to use the funds to help start minority businesses in underdeveloped areas. Also, some St. James citizens, including the local NAACP chapter president, formed a non-profit job training center with $500,000 from Shintech itself. So, the state of Louisiana was kicking in money to develop minority businesses, and Shintech was putting its money on local businesses. I have to ask a question ... wasn't that the ultimate goal of the push for environmental justice? I believe the answer is yes.

Apparently though, you can't take a fight away from people who really want to be in one. Despite the money pouring into the community, Greenpeace went ahead and filed a court motion under the EPA's environmental racism rules (Title VI) to stop the Louisiana Department of Environmental Quality from issuing an operating permit to Shintech. The DEQ held a two-day public hearing in January 1998 in the gymnasium of an elementary school in nearby Romeville, which allowed residents and non-residents alike many hours to shout, scream, pray, chastise, preach, read prose and recite poetry - some more than once. Consistent with her past calls for clarity, Marylee Orr of LEAN issued an intelligent, concise call for a clear understanding of the issues wrapped up in the term environmental justice and a true determination of what was fair, then offered the crowd a large cake in the shape of Shintech so everyone could eat the company in effigy.

In the midst of that meeting, however, Ernest Johnson, president of the Louisiana NAACP, stood before the microphone to issue a report supported by data gathered years earlier by LSU, Tulane and Louisiana Health Department researchers. He said minority residents of Romeville and Convent were not dying of cancer caused by pollution. They were dying because they couldn't afford health care. So the real issue wasn't potential pollution from Shintech; the real issue was poverty, a conclusion echoed by the state health department, cancer specialists and researchers at Loyola and Tulane Universities. And poverty could be offset by industrial development such as the Shintech plans. In one short paragraph, Johnson removed the state NAACP from the ambiguous environmental equity track and placed it firmly back on the civil rights track.

Talk about a monkey wrench in the draw works. National civil rights and environmental equity advocates had just been blind sided by the people they were fighting to protect. Why? Because local, grassroots environmental leaders and participants had witnessed their agenda and their voices squelched by outside groups with other more global, more radical agendas. The national NAACP and the Rev. Jesse Jackson could afford to shut down one more insignificant plant to further their cause. The residents of St. James Parish and their local minority leaders, however, could not. They didn't regard the Shintech employment and development opportunities as insignificant. They saw it as an opportunity for jobs and community enhancement. For a while, it seemed Shintech might have a fighting chance to open its doors and fill its employment rolls. However, the fight was bigger than the company. Shintech had become a national object. National civil rights organizations, environmental groups, politicians, the EPA and others regarded the company's quest for permits as the first real test case of applying civil rights laws to industrial development. Not needing that kind of attention or notoriety, Shintech announced in September, 1998 that it would not open its plant in St. James Parish, but would instead open a smaller facility upriver in Plaquemine, Louisiana. According to Shintech Controller Dick Mason, the new site would raise fewer environmental justice questions because the town was more affluent and more white.

That move by Shintech caused a huge dilemma for radical enviros and civil rights leaders - what to do? They had been planning for a fight in the courthouse to finally work out the understanding of the new environmental equity rules, but now they were left without their test case. It wasn't exactly the victory they were hoping for, so the search for the next test case began immediately. The civil rights crowd quickly shined their environmental equity spotlight on Select Steel Corporation of America near Flint, Michigan. That company had earlier applied for permits to open a steel recycling mill but immediately drew new fire as the next best test case. That case didn't end in victory for the civil rights crowd either, by the way. The next month, the EPA made a significant ruling, probably the first to help clear up these rather murky waters. The agency said because emissions from existing and future projected operations at Select Steel didn't exceed EPA standards, minorities could not claim they were affected adversely or in greater proportions than whites. The new ruling said pretty clearly that minorities must first show harm, in this case emissions which exceed national pollution standards. Then and only then could minorities proceed with a claim that the harm to them was greater than to whites.

Slowed, but not stopped, the national civil rights crowd continued its search for cases to prove environmental injustice, nationalizing any local issue which met the proper criteria. Yes, local groups with nifty acronyms sprang up to fight industrial expansion, but they were always joined, bolstered and encouraged by the national groups. In the Shintech case, a local group called AWARE (Alliance against Waste and Action to Restore the Environment) emerged on the scene backed up by Greenpeace, the Tulane Environmental Law Clinic, the EPA office of Civil Rights and even Jesse Jackson. In fact, another group, the St. James Citizens for Jobs and the Environment, became little more than a front for Greenpeace press releases until the local newspaper caught on to this and began to attribute quotes from the group to both St. James Citizens and Greenpeace.

Let me state here and now that I believe the issues wrapped up in the terms environmental equity, racism and justice are real. They are important, and I think they should be discussed and resolved. However, I also believe they completely cloud the issues of clean air, land and water, which were the original goals of the grassroots environmental movement. When the civil rights crowd stole the agenda from the grassroots movement just to capture its members and support, the environmental movement took a huge hit. But agenda stealing alone was only part of the hit. The introduction of radical, militant civil rights tactics also hurt the grassroots movement. The PBS television series "Eyes on the Prize," produced by Henry Hampton and written by Steve Fayer, outlined the history and advances of the civil rights movement and was a great case study of civil disobedience and social tactics for change. In it, one young black leader stated they often didn't know the source of power in one town or another until they "started stomping around... stepped on a toe... and heard someone holler." That was how the activists found their true targets.

Although, the tactics of disruption, disobedience, open confrontation, shouting and name-calling were new to the moderate grass roots environmentalists, they were, of course, old, familiar weapons to the civil rights crowd. Moderates simply believed those militant tactics would create more heat than light and wanted instead to sit down with politicians, industrialists and other citizens to work out solutions in a civil way.

Enter Pat Bryant, head of the Gulf Coast Tenants Association of New Orleans. He once told me, "These folks who waste time eating lunch (with industrial management) and talking are wasting time and killing people."

Bryant's group was composed of minority citizens of New Orleans who were not property owners or homeowners. The tenants association had been around long before environmental issues were taken up by minorities, fighting for basic housing rights for non-homeowners and under-served minorities (to use some civil rights lingo).

I interviewed Bryant in 1992. "Had someone told me 10 years ago I would have been working in environmental concerns, I would have laughed in their face," he said, " We have set ourselves to be experts in this area alone. Our leaders said in 1983 environmental concerns were important. Some people had no potable water ... some thought they were eating contaminated fish."

Bryant swung his group straight into the environmental movement along the equity, justice and racism lines. He once told me, "I'm an agitator (chuckle). I stir things up. As we agitate and organize, our goal is to create a new mass of which we are the center of gravity." Well, surprise, surprise. That was exactly the civil rights tactic voiced so well by the young civil rights leader in "Eyes on the Prize," but which eventually blurred the real goal of cleaning up pollution. Nevertheless, the tenants association and Bryant received more press, more publicity and more action after joining up with the enviros than they had ever received before or since. That's pretty much the way most civil rights groups have gotten ink and air time lately, which further illustrates my point that the environmental movement has become a tool used by civil rights organizations gain the attention they need to advance their own goals.

By the late 90s, the predominantly moderate grassroots environmental movement was flooded by civil rights activists hoping to ride its coattails of success, corral a rather large body of supporters and donors, and advance their separate agenda of forcing industry to pay to make life better for minorities who lived in and around industrial sites. At first, environmentalists welcomed the new voices, energy and support. But when the alternate agendas became clear, many moderates, even environmental activists, bailed out. The movement became tainted with murky issues which didn't necessarily lead to cleaning up the environment. Now it seems that no environmental issue is raised anywhere without someone screaming against environmental racism or for environmental justice.

The problem is that the terms environmental racism, equity and justice are still very unclear, so it is difficult at best to know exactly when we've reached a state of no environmental racism or of complete environmental justice. Although the Environmental Protection Agency is under orders to consider environmental justice/equity/racism when granting permits for new construction at plant sites, it has not clearly defined the terms. Worse, the EPA has instructed state environmental agencies to consider environmental equity as well, but has done little to clarify the issues states should consider or explain exactly how to do so. My personal opinion is that the oft and rather loosely used terms of environmental racism/equity/justice are now simply buzz words, used by politicians, minority leaders and activists alike to evoke action from those people and organizations who do not wish to be called racist. President Clinton slings the term environmental equity around conveniently to pronounce new rules and gain minority support, but the word is beginning to ring rather hollow due to overuse.

In a small step to clarify the matter, the EPA issued interim guidelines in early 1998, detailing how it would use Title VI of the Civil Rights Act to determine environmental equity issues. Those guidelines were little or no help to states, particularly portions of the guidelines which dealt with mitigating circumstances. Apparently it is going to be impossible for states, cities, residents and corporations to agree on what exactly will mitigate a complaint, or what constitutes a "minority population" or "low-income population" as stated in Clinton's executive order. This presents a huge problem for state agencies trying to seek environmental equity. Title VI clearly prohibits recipients of federal funds - like state environmental agencies - to discriminate on the basis of race, color or national origin. So states could lose federal funds if they don't comply, but compliance is nearly impossible.

"None of this is defined. We in this state and those in other states have no clear definitions." That's Jim Friloux of the Louisiana Department of Environmental Quality, the state agency given the task of following the federal mandate to "consider" environmental equity issues when handing out industrial permits. "That's the criticism of the federal law...no definitions." Friloux explained that although citizen groups across the country were trying to use the Civil Rights Act of 1964 - as per EPA guidelines - to impose their will on local industry, the law in no way addresses industrial development, jobs or pollution. It is a strategy destined for failure, and doomed to create further animosity between citizens and industrialists and between environmental radicals and moderates.

One of the most noted critics of the new environmental equity rules is former EPA counsel Gerald H. Yamada. He served on the federal Interagency Working Group which was formed to determine how to implement Clinton's executive order. He left the EPA a year after that order was issued and has been vocal in the media, stating the EPA is doing it all wrong. During a week-long meeting of the National Environmental Justice Advisory Council (NEJAC) held in Baton Rouge, Louisiana in late fall 1998, Yamada said, "This is a model study of how not to implement a new program. EPA is making this up as they go along."

Joining Yamada in criticism is political scientist Christopher Foreman, author of *The Promise and Peril of Environmental Justice* (Brookings Institution, 1998). Foreman describes himself as a black liberal Democrat who voted for President Clinton twice. He began research for his book shortly after President Clinton issued his executive order calling for environmental justice and quickly became concerned that the whole issue "was heading in the wrong direction." Foreman's main argument is that environmental justice proceedings evoke a lot of emotion, anger and even outrage among minority communities, but is based on poor science. He believes the enormous amount of energy spent fighting individual toxic sites leaves citizens with few emotional and physical reserves to devote to more important public health issues. He also believes the focus of radical environmental organizations on health risks from toxic sites is misplaced compared to the more pressing issues among minorities including teenage smoking, low birth weights, poor prenatal care, drug abuse, heart disease, violence, obesity and AIDS, among others. Calling early research on environmental justice "unsophisticated at best" and "junk at worst," Foreman says the main flaw in the environmental justice crowd's argument is their inability to define a minority community or to clearly articulate the desired end result of environmental justice policy.

Late in 1998, the U.S. Environmental Protection Agency's environmental justice officials met with the Mississippi River Corridor Task Force, a group formed following the Shintech controversy, with the directive of examining the economic and environmental interests of locating new plants along the Mississippi River between

Baton Rouge and New Orleans. An award-winning environmental reporter, Mike Dunne, was prompted to write a column following those discussions in which he stated, "There were plenty of questions but few answers." He highlights the comments of Michael Mattheisen of EPA's Office of Civil Rights who said the definitions of "community" (as in minority community) and "disparate burden" (as in those minority communities bear a disparate burden of industrial pollution) would be hammered out in litigation (as in long, extended lawsuits where federal judges are allowed to make, rather than interpret law.) Mattheisen also stated other elements of the law would be decided by the "public participation process" or might be "in the discretion and judgment of EPA." In his column, Dunne wrote, "That is not a clear answer." None of this is very reassuring.

Time and again, those I've spoken to and those who have spoken out say the same thing; environmental justice is a vague notion, difficult to enforce, hyped with anger and emotion and backed up by little science. Worse, environmental justice has little to do with the environment and everything to do with a new, antagonistic agenda for civil rights. Foreman believes environmental justice activists place too much emphasis on the health risks of toxic sites and industrial plants, even though not a shred of direct evidence indicates harm. The emphasis on health risk, though, is a well-planned strategy. According to Dr. LuAnn White, public health researcher at Tulane University and keen observer of the environmental movement, the claim of health risks is an important part of the litigation process. Yes, citizens could file suit against one company or another under existing pollution laws to press for clean-up or pollution reduction. However, those laws don't allow the enormous sums of personal pay-offs as do suits filed under personal liability laws. In short, if a resident claims harm to health from pollution, the laws allow huge monetary awards if liability is proven. Ever wonder why practically every news story you see or read about pollution contains several choice quotes from locals claiming they got sick? That's where the money is. Environmental ideals and the quest for clean air don't enter the picture, personal reward does. So what the environmental movement has come down to is a series of battles against single sites or companies where residents and radicals alike claim harm to health and seek millions of dollars in retribution, not clean air.

But radicals and civil rights leaders aren't naive, they realize proving harm to specific individuals is difficult, if not impossible. But, they also realize corporations are often reluctant to throw their fate onto the mercy of the court and may wish to settle lawsuits or avoid them all together. Enviro-pros encourage local citizens to seek many things from corporations willing to settle differences out of court - new roads, new schools, public transportation or even the ability to sell their house to the company for an inflated amount and be moved to a new home. In the era of environmental justice, large, industrial companies have become milk cows to "correct" injustice to local minority communities however vague the definitions of either "environmental justice" or "minority community."

The pressure on states to act on such vagaries finally became too great, and in late 1998, an organization made up of state environmental regulators called for a halt. The Environmental Council of States met in New Orleans on March 26, 1998 and issued a resolution labeling the EPA's policy of using the civil rights law to enforce environmental justice as "unworkable." One of the things the resolution asks for is the use of "precise, scientific and peer-reviewed methods" to determine health risks.

Undaunted, the Clinton administration stands by its mandate for all federal agencies to "consider" environmental equity in its permitting, and the EPA stands by its insistence that states do likewise, despite the lack of definition. At this writing, the EPA had 15 environmental justice cases pending across the nation, searching for the best one to test its new civil rights mandates. That means states are scrambling to determine for themselves what equity means. In Louisiana, the state environmental quality department went so far as to hire former chairman of the U.S. Commission of Civil Rights, Arthur A. Fletcher as a consultant to help understand the law. Although the state paid him nearly $50,000 for a year-long contract, he sounded more like an advisor to local civil rights leaders than a consultant to the agency. If anyone was uncertain of the link between civil rights and environmental equity before his arrival, his first remarks to a state task force made it crystal clear. He flatly stated to the task force that neighborhood economic development was, "the final push of the civil rights movement." Then he used the platform to urge communities to use the issue of environmental equity as a "tool" to achieve economic development. Apparently the great state of Louisiana wanted help to understand the EPA ruling, but instead succeeded only in getting an extreme civil rights activist placed on the payroll. Fletcher made it clear he would spend his time telling chemical companies what their policies were doing to the people of St. James Parish. Nearly a year after Fletcher was hired, the state remained in doubt about exactly what environmental justice and equity were. At the ime of this writing, state officials were extremely concerned that since Louisiana was regarded by many as a test site for the new environmental justice rules, normal business development would be put on hold indefinitely until the whole mess was sorted out. And they feared any ruling at all from the EPA was bound to be a detriment. "It (the federal mandate) is so vague and ill defined - it's dangerous as we see it. It can completely halt industrial growth," Jim Friloux of the Department of Environmental Quality told me.

That may be, but where many conservatives see danger is in the willingness of the President and the federal government to issue such sweeping, poorly defined mandates based on poor or no data. To this day, there is still no good evidence to suggest minorities are intentionally targeted by industry to breath bad air, unless civil rights leaders are privy to a whole new method of physics and meteorology. Stop and think about this argument made by the environmental equity crowd to see if it fits conventional science. Environmental equity advocates claim that industrial development in rural, mostly poor minority communities is bad. Their reason is that most of the new industrial jobs go to outsiders and management jobs go mainly to whites while the surrounding poor black communities are unjustly subject to the resulting industrial pollution. Okay, here's where the new physics and meteorology must come in, so let me be the first to raise my hand and ask: What manner of unusual air currents or physical occurrences causes the industrial pollution made by the white managers and employees inside the plant fence to affect only the poor blacks on the outside? The answer, of course, is there are none. Despite the argument that local residents are unjustly subject to pollution while managers and employees escape all harm, no data supports this conclusion. It's a silly notion and, ironically, the whole idea of selective harm to surrounding minorities by pollution made inside the gates by white, outsiders was in direct opposition to the assertions of trade unions in the environmental movement. Those unions argued for years that industrial pollution did in fact harm workers inside the plants as well as the residents outside.

But the selective harm of industrial pollution is just one example of fuzzy thinking and selective use of data prevalent among those who wish to forward their pseudo environmental agendas. Consider another example of selective thinking. When Ben Chavis, the relentless, hard driving advocate of civil rights, held aloft the first "report" to use the term environmental racism in order to shine the first bright light on the subject, the civil rights community upheld his report as "truths." Yet, that report was one he wrote and was issued by the civil rights commission he led. Not exactly a scenario for unbiased reporting. Yet when the Louisiana NAACP leader held up a Louisiana State University report, which he was not a part of and which he did not write, showing cancer rates among blacks in south Louisiana were not out of line with the rest of the nation, he was shouted down as being pro-white industry by the environmental equity crowd, including other civil rights leaders. That brand of selective thinking and the unjustified, instant, angry reaction by the enviro equity crowd in the face of apparently legitimate data is what drove moderates away by the thousands.

Keep in mind, please, that I believe the issues wrapped up in the terms environmental racism, equity and justice are quite real and deserve discussion and resolution, but they are not, strictly speaking, environmental issues. They are social issues. That blurring of environmental problems with social and civil rights issues, coupled with the instant anger evoked by environmental equity leaders has severely hurt a grassroots movement which relied heavily on a largely moderate membership to wage a war against pollution. Historically, for whatever reason, a greater number of moderate citizens were willing to join a clear fight against dirty air and water than to get in line with some vague claim against industrial America of injustice. To some, environmental justice tactics smacked of race baiting; no wonder the grassroots environmental movement waned after the introduction of the term. It represented little more than a re-birth of the civil rights movement and appeared to be fueled by the same people and organizations who have always been civil rights advocates. So when the goals of the environmental movement got blurry and the tactics turned mean-spirited, grassroots grannies, enviro moms and dads and school children who joined the movement by the thousands left the building. The environmental racism/equity/justice issues are just too unclear, overused and shouted too frequently in anger by people with other agendas. The real enviros with their clear, simple agenda for clean air and water were no longer in control and decided that it was time to go. Still worse, many were pushed out. Remember Janice Dickerson, the woman working for environmental equity in Louisiana? Her frankness and openness make her instantly likeable, but her comments to me were crystal clear and unwavering. I (as a white man) and other white people were not welcome in her neighborhood, or any other predominately black neighborhood, to fight for environmental justice unless we were specifically invited. Her belief was she and her neighbors and members of communities like hers needed no hand-holding or guidance or "looking after" by outside organizations, especially predominately white ones. That's fine. I told her I respected that opinion. The problem for the environmental movement, though, was her logic was exclusive. No wonder moderates dropped out of the movement. The main movers and shakers in the 90s were the civil rights advocates, and they were specifically telling white moderates they were unwelcome until invited. The trouble was, no invitations were forthcoming. Moderates dropped out in droves when they saw the environmental movement taken away from them, creating perhaps the greatest blow to that once simple, effective grassroots movement.

# Global Green - A New Social Order

One of the least recognized agendas of some members of the environmental movement is to realize their vision of a new social order by raising the protection of the environment above the needs of humanity. The enviros who want this are beyond liberal or even radical. I'll call them the ultra radicals - or ultra greens. Earth First!, the most radical elements of Greenpeace, The Green Party, PETA (People for the Ethical Treatment of Animals) and numerous individuals across the nation and the globe fit into this group. The new social order they are striving for would elevate animal and plant rights to a constitutionally protected state - if indeed these folks believed in the constitution. On its face, the new order would be a socialistic government which would greatly reduce human rights and activities in order to ensure a better environment for other living things. As I read and listen to these people, squads of roving environmental police, looking to cite, arrest or jail environmental offenders would not be a far fetched idea within their social order. I argue the green agenda is simply a mechanism to attain a form of totalitarianism for the new millennium. The thought of that kind of massive, government enforced daily lifestyle terrifies me, and I believe moderate environmentalists would agree. Moderates deplore the tactics, strategy and agenda of their ultra-green counterparts and don't want to be associated with them. Believe it or not, most moderates want to improve the environment without breaking the law, without destroying private property and without hurting anyone.

Many authors have tried to shine a light on this subject. An anthology compiled by Peter C. List entitled *Radical Environmentalism - Philosophy and Tactics* (Wadsworth Publishing Company, Belmont, Ca.) details exactly what these ultra radicals have said about their agenda. You'll learn that these folks are serious about creating a society which places animals and plants above humankind and are working the political process to make it happen. If you like the idea of socialism, totalitarianism and even communism coupled with an environmental agenda, you'll love these guys and gals. And you'll come away with a whole new vocabulary too - ecofeminism, ecotage, deep ecology, ecotage and the like. I also encourage you to peruse the internet. The internet is a chief tool of environmental extremists, so just start some searches under the terms mentioned and you'll be able to read exactly what they are espousing. You'll also find moderates providing some very good, critical analyses of the extremists' agenda to create a society where government restricts individual human rights and freedoms whenever they impact animal and plant activity. Folks, I'm all for clean air, land and water and responsible use of resources (plants and animals), but read a couple of the following quotes and see how far flung these people are. It is no stretch to envision that in their world, the enviro police will show up to cuff you and seize your house because you cut the grass.

In Guelph, Ontario, during a public meeting to develop an environmental protection plan for the city, the true agenda of the radical greens was revealed when someone in the audience suggested the planning committee have a member to represent social issues. When queried, the audience member stated that this person would represent human rights, poverty, low-income housing and animal rights. It was clear the radicals wanted the government panel to use the environment to reach into most aspects of life. The panel, for the moment, tabled the idea. (from EnviroScan # 136)

I argue that the tabling is only temporary. The one consistent feature of the ultras is they don't give up. They will continue to suggest, move, vote for or by any other means strive for that animal rights representative until their suggestions don't sound so unusual anymore. The board may give in, and others would probably follow. So the good citizens of Guelph will, by decree of their government, be facing issues of which animal (and presumably plant) "rights" will be a legitimate component. Here's more ...

"We already have too much economic growth in the United States. Economic growth in rich countries like ours is the disease, not the cure." - Paul Erlich, Stanford University biologist and advisor to Vice President Al Gore

"I think if we don't overthrow capitalism, we don't have a chance of saving the world ecologically. I think it is possible to have an ecological society under socialism." - Judi Barri of Earth First! in Walter Williams column, Heritage Features Syndicate, June 25, 1992

"Capitalism is a cancer in the biosphere." - Steve Chase, ed., *Defending the Earth: A Dialogue Between Murray Bookchin and Dave Foreman*, Boston South End Press, 1991.

"Childbearing [should be] a punishable crime against society, unless the parents hold a government license...All potential parents [should be] required to use contraceptive chemicals, the government issuing antidotes to citizens chosen for childbearing." - Dave Brower, Friends of the Earth.

"We in the Green movement, aspire to a cultural model in which the killing of a forest will be considered more contemptible and more criminal than the sale of 6-year-old children to Asian brothels." - Carl Amery, Green Party of West Germany.

These are just a few nuggets gathered at various green sites on the internet. But, the most extreme views are published in more traditional sources as well, like books written by the extremists themselves. Nat Hentoff, in his newspaper column, tried to point out the mindset of one of the most unbelievably anti-human writers now teaching our country's future leaders - Peter Singer, Princeton professor. He stated in his book *Practical Ethics,* "Human babies are not born self-aware, or capable of grasping that they exist over time. They are not persons." Singer, also author *of Should the Baby Live?,* wrote that because animals are self aware and newborns are not, "the life of a newborn is of less value than the life of a pig, a dog or a chimpanzee."

And finally, words from the man who led the way In placing animals above humans, Cleveland Amory. The former editor of The Saturday Evening Post, radio personality, and best selling author known to many, was also an animal rights advocate and founder of The Fund for Animals. When he passed away in 1998, National Public Radio replayed an old interview where he stated, "If we are God's chosen species, he chose poorly. There are lots of animals more deserving of the hereafter than we are."

Scared yet? Moderates in the movement were, and they got out.

The key to the ultra green enviro social strategy appears to be the fact that motivating government to regulate business and industry by elevating plant and animal rights over human rights is much easier when the environment appears to be in real jeopardy. So the radical anti-business, anti-technology, anti-human rights crowd played the enviro card with each hand. After all, who could argue *against* clean air, land or water or *for* unrestricted deforestation or *for* driving animals into extinction? Even the political conservatives, who usually favored business development, were jumping on the enviro wagon, calling for greater accountability for pollution, more specific information from polluters on what they were actually sending into the environment, bigger spending on pollution control devices in industry and restrictions on individual human activities which appeared to pollute. But the final regulations, actions and restrictions of business activity posed by the extreme greens would dramatically change our society. In the extreme green world, humans would kill no animals or plants for any reason. Businesses which make goods from natural resources like oil, gas, wood or minerals would be closed, and we would revert to an agrarian, tribe-like culture, using up few or no non-renewable resources. That folks, is what I mean by extreme.

Ultra greens' writings, speeches and actions are based on the belief that a higher law exists than those we currently have on the books of society. That higher law is to protect all living things by regarding them equally or even in a superior position to humans. The extreme green tactics of civil disorder and disobedience are born in the ultras' belief that democracy is too slow, the masses are too ignorant and a philosophy of civil liberties and human rights is misguided. The extremes would rather subvert the law than attempt to change it. In fact, they coined a whole new word to describe what they do - ecotage. Ecotage is sabotage which is designed to protect the environment, and practitioners say they are operating under a higher law than man's law - a law to protect the earth. Monkeywrenching is one style made famous in ultra green literature; it is the practice of dismantling or disabling construction equipment. You can learn specific methods in *Ecodefense: A Field Guide to Monkeywrenching,* 3rd ed. (Dave Foreman and Bill Haygood). Or from Cafe Underground (www.cafeunderground.com), where a another complete manual exists. Here's part of the introduction:

"I have to admit that the height of my career has been sabotaging dumpers, JCBs and 2 Caterpillars within fifty yards of the security guard's office. The worst was certainly sabotaging some road construction equipment and having to walk five miles home covered in diesel. In between I've scratched Porches, 'improved' billboards, sugared fuel stores and glued locks." - from Ozymandia's Sabotage Handbook

There is also an organization in the fringes of the environmental movement called The Luddites who profess to want to ignore, remove or destroy any technology, longing instead for the days of horses, thatched roofs and subsistence farming. Greenpeace regular troops are actually rather tame in comparison to the ultras. Greenpeacers may throw themselves on top of baby seals, in front of harpoons, into nuclear test sites and in front of the discharge pipes of industry. And they believe in social disobedience, also known to some as trespassing, vandalism and disturbing the peace. However, the crowd of ultra-radicals over at Earth First! and within other ultra groups look on all of that as child's play. These are the folks who don't just throw themselves down in front of bulldozers or chain themselves to corporate gates. Instead the ultra radicals, who believe they are following a higher calling into a literal war on society, would rather sneak in at night and drive huge steel spikes deep into trees, so the next morning unknowing loggers' chain saws jam, twist and break, putting the safety of the startled loggers at great risk. Ultra greens also like the idea of sneaking into the vehicle pool and dumping sand into crankcases of the bulldozers, trucks and tractors and destroying as much property as they can. And some ultras don't hesitate to use more violent and potentially more deadly tactics which threaten the safety of innocent citizens. Calling themselves the Earth Night Action Group, one ultra faction cut off power to a part of Santa Cruz to protest participation in Earth Day 1990 by Pacific Gas and Electric Company. That put 90,000 residents in the dark, including any hospitals, doctor's offices, fire and police stations in the area.

Yet those are only the readily visible elements of the radical greens. There is a larger, unseen effort to alter society through an environmental "do anything to protect the earth and the animals" agenda. In the ultra vision of the future, we will be set back many decades in technology and lifestyle. Unlikely you say? Well let me tell you, I've met folks who prefer horses to cars, grass huts to houses and free ranging animals to free ranging people, and their views are slowly but methodically easing into society and into government regulation. The ultras know how to win very tiny victories over long periods of time, and they know how to keep constant pressure on the political process in a way that we all get used to it. I call this nearly invisible process extreme creep.

Here's how extreme creep has already gained ground. To fulfill their wish for animal rights over human rights, ultras first worked to ban the hunting of rare and endangered animals. Not much resistance there; that was an easy first call, even for moderates. The animals weren't plentiful enough to be a real source of food or clothing, and humans weren't barbarians after all, so why not design an endangered species list to try to protect them? You could almost hear Yul Brenner in his line from The Ten Commandments: "So let it be written, so let it be done." Bald eagles, Asian tigers, and other exotic animals could then live and breathe more safely; they were officially protected by the endangered species list. Next step was to protect whales, and banning whaling was another relatively easy, global call. Sure, it took a few years for Japan and the USSR to get on board, but it finally happened. Few people could see the real need for whaling when substitutes for nearly every part of the whale were available elsewhere, and those pesky Japanese and Soviets could find another line of work couldn't they? Okay, Greenpeace won that one. No more whaling and no more need to risk human life by driving Greenpeace inflatable speedboats into the sights of harpoon cannons.

Banning the hunting of baby seals was a little more difficult, but we took the step because we had already done so with exotics and whales, and the move seemed logical. Baby seals offered few things we really needed or couldn't get elsewhere, and thanks again to Greenpeace, we all saw the mothers look mournfully into the rolling cameras as the hunters hit the little creatures on the head. After all, who wanted a fur coat whose previous owner was clubbed to death in front of its mom? Sure, some Nova Scotians would be out of work, but they were a hardy lot, they could find something else to do, and if not, they could move, right? So, ban the killing of baby seals we did. The question became what ban would be next? Elephants were rare and so was the black rhinoceros. Banning their hunting seemed a good idea, so we did that too. So far so good. Okay, if the whale ban was successful because the whale provided nothing we couldn't find elsewhere, how about the fur trade in general? Since fake fur was readily available, the slaughter of fur bearing animals seemed superfluous. How about a ban on the hunting of animals strictly for fur? All right, done.

Now for a look at the future and where extreme creep could lead us. Since we've already established fake fur is readily available and hunting wild animals for fur is bad, why not also say that killing farm or pen raised animals for fur is bad and ban that too? And if killing animals for fur is bad, isn't it worse to kill them to eat them? That seems a bit barbaric when we can grow corn. So let's ban hunting and raising animals for use as food. Believe it or not, that argument actually seems plausible to many when couched in the right terms. We as a society have endowed large, furry animals with human characteristics in literature, TV and movies for a long time, so maybe pulling the Remington off the rack and lining up the cross hairs on Bambi, Bessie and Yogi isn't such a good idea after all. But what about all those fish in the sea? They certainly don't have feelings, or do they?

As a society, we have already banned killing dolphins, whales and sea turtles, but of course they are different than fish. Dolphins and whales are mammals, and everyone knows Flipper has feelings. And sea turtles are magnificent creatures with a fascinating, partially land-based lifestyle. But what about all the lesser sea creatures? We're not going to worry so much about them now, are we? Not so fast. Some crustacean's rights folks have already stormed grocery stores in the northeast, demanding that startled owners free the live lobsters in their display tanks. Most savvy grocers said something like, "Fine, that'll be $8.95 a pound, and you can let 'em go wherever you want." Now let me ask you, is that news? You bet – as a rather quirky feature story about a small group of lobster lovers doing that crazy Born Free thing. However, rather than treat the lobster freeing as an unusual, almost fluffy feature story, that great institution known as network news actually granted it serious coverage, which gave the lobster freeing nuts some credibility. Millions of TV viewers watched as people with very little else to do bought some live lobsters, removed those painful little rubber bands from around their claws and set them free at last into the Atlantic Ocean. Does the heart good, doesn't it? I wonder if it mattered to the lobster rights folks that they had just taken lobsters from relative peace and quiet and released them into perhaps the most lethal, most dangerous environment imaginable - the wide open ocean. I suppose we should debate which is a better death for a lobster, being boiled alive or torn in half by a shark?

Now let me ask you this. Those lobster loving folks stood face to face with that grocer - who was simply a private citizen and businessman - and demanded he give up part of his livelihood just because they thought he should. Do you think they would pass a law to do the very same thing if they could? You bet they would. And as far-fetched as that idea may seem to you, not far behind it is the notion that plants also may be endowed with feelings and therefore certain inalienable rights. They'll need laws to protect them too. By now, I'm sure you're thinking, "Yeah, right, I really believe people will actually fight for plants' rights." Ever been to Athens, Georgia? A former University of Georgia Professor who lived there loved a tree in his yard so much, he actually willed it ownership of the ground it grew on. After he died, the tree received a stone marker quoting the deed filed in the Clarke County court, which reads...

"... for and consideration of the great love I bear this tree and the great desire I have for its protection for all time, I convey entire possession of itself and all land within eight feet of the tree on all sides. - William H. Jackson."

Apparently until the tree dies, no one, not even the homeowners on the lot where it resides can have use of that patch of land.

That one little paragraph is how environmental creep starts, and Athens is a good example. Is the tree protected by the vigilance of Greenpeace or Earth First! Nope. Is it monitored by extremists or movers and shakers for a new social order? Nope. The tree is tended to and cared for by the very extremist, left wing, ultra green group that calls itself the Junior Ladies Garden Club! And that is the result of enviro-creep; even the most moderate, civil and civic minded organization with the best intentions (sorry ladies, I didn't really mean all that ultra-green stuff) can fall trap to enviro-

creep under the right circumstances. But there's more. What began only as a highly unusual, perhaps even controversial last will and testament became ingrained in Athens society. The Junior Ladies Garden Club began to care for the tree and erected the stone marker. Then George Foster Peabody himself (who's name is on one of the highest national journalism awards available - the Peabody Award) donated money to build a granite post and chain fence for the tree. But here's the real news. The tree within that fence is not the original tree, which blew over in a storm on October 9, 1942. The tree standing today is an offspring, raised from an acorn of the original and planted by the Junior Ladies Garden Club on October 9, 1946. Yet, according to literature from the Athens Convention and Visitors Bureau,
"the Athens community recognizes the tree's title to the surrounding land and takes added measures to protect Jackson's favorite oak, the deed has never been tested in the courts."

Suppose I tried to do that same thing in another town today. Let's say I owned a lot downtown, next to a street the city wanted to widen. But I willed the lot to my tree - or rose bush - and died a week before the city was to begin work. Could the city tear down the tree or the bush? Would it? I argue that conventional community values would prevail, and the tree would go, particularly if the street being widened served a hospital. Anyone throwing themselves in front of my beloved rose bush to stop the bulldozers would be regarded as extreme. But not in Athens, Georgia where "the community recognizes the tree's title."

All right, maybe I'm making too much noise about that one tree in Athens. But just suppose a nice young couple were to move into the house on that lot and give birth to a child that needs special attention of some kind, or perhaps the folks that live there now have an accident and need special equipment - a wide driveway for a vehicle equipped with a lift, or a special room built onto the house, or anything that would require use of the corner of that lot. Guess what. Apparently, the tree would have rights over the needs of the child or the residents. I stood next to that tree in 1981, and let me tell you it was nice, but no nicer than any other tree in Athens. I would cut it down in a second to provide for the needs of a small child, and plant one of its acorns somewhere else. But I pity the Athens homeowners who are the first to challenge the will and tear down the stand-in tree. I can see the news story now: "Cold-hearted tree killers fight Ladies Garden Club." Based on what I've seen, the real extremists in the grassroots movement would shriek in horror and surround the tree with their bodies, sabotage the bulldozer or spike the tree to cripple chainsaws. The vision of garden clubbers accepting such hypothetical help demonstrates a perfect example of enviro-creep.

Athens, Georgia aside, that extreme mentality is what scared off the huge moderate membership in the grassroots environmental movement. That kind of radicalism and extremism didn't create the huge crowds in the movement; school plays on recycling, enviro moms for clean air and Grandmoms with sick babies did. The real anti-business, anti-technology and even anti-human sentiment belongs to an extreme minority in the environmental movement which simply harnessed the moderates' genuine concern about protecting the planet to try to engineer social changes which actually diminish human rights to a level below the rights of animals and even plants. The extreme tactic is not one of large, dramatic steps, but of many, small almost indiscernible ones; the Tree That Owns Itself, the suggestion at a town meeting to consider animal rights, a persistent disobedience of leash laws, subtle but continual pressure in literature and film to depict animal feelings, rights and struggles.

Enviro-creep started with the simple recognition of extreme positions as actual concerns within the environmental grassroots movement, which was growing in membership and integrity. Surrounded by a large body of concerned citizens, including conservatives and moderates, those voices for societal change sounded plausible at first. Then, as their voices grew stronger and their initial, rather small suggested changes took place, the extreme element spoke out a little louder and a little more often to institute a few more changes. Soon, politicians, media and other environmental members began to listen to the extremists and engage them in conversation and policy-making. Before long, extremists gained what they wanted: recognition as legitimate and thoughtful members of the movement. That's the success strategy of extreme creep. It's hard to kick someone out of the room for an extreme idea when you've already listened to and acted on their earlier, less extreme ideas

Now for an example of the ultra extremism I'm talking about right out of today's headlines: environmental justice. Remember the drift - or creep - we've already discovered? Environmental racism begat environmental equity which begat environmental justice. That change in name accurately reflected a change in strategy as well; the strategy of placing industrial development in direct opposition to civil justice. Some recent remarks by those leading the environmental justice movement are quite revealing as is the pattern of creep revealed by them.

Tulane University opened its Environmental Law Clinic in 1986 to represent minorities in south Louisiana who could not afford legal help to fight environmental racism. Led by Bob Kuehn, the clinic became somewhat famous, as Kuehn and Tulane law students led the legal battle whenever a big company appeared to be dumping pollution on poor black people. That seemed okay at first. Someone had to represent the poor, and university law clinics in general have had sterling reputations for doing just that. However, the Tulane Environmental Law Clinic used extreme creep to its advantage. In not too many years, the clinic wasn't representing individual poor people, but entire communities and environmental organizations. Any time someone shouted environmental racism, equity or justice, the clinic was there, and I can assure you it was not to clean up the air. The clinic appeared to be pressing for no less than a new social order, where industrialism was decried as murder. The clinic did not strive for clean air as much as it fought capitalism and industrialism at every turn, in the name of environmental justice.

Ironically, the premise of environmental justice is to spread the industrial wealth across all racial and economic boundaries. Companies which pollute should not take their toll on minorities while the majority gets the payroll. In most environmental justice rallies I attended, many, many citizens called not for the closing of a given plant, but for jobs from it, equitable taxation or municipal fees to invest in local roads, schools and facilities, and for employees and managers to invest themselves and their money in local commerce and philanthropy. In short, environmental justice, as it has been explained to me, relies on industrial growth to fuel local, minority economies while protecting the environment.

Environmentalists and the Tulane clinic didn't see things that way. Yes, they argued for some of these economic boosts in the Shintech case mentioned earlier, as well as argued against the potentially polluting activities of Shintech. But if the law clinic were true to the principal of environmental justice, it would have recognized and welcomed the money Shintech invested in local economic development ventures and equally welcomed the state of Louisiana's investment through the department of economic development. But Kuehn and the clinic did not. They labeled investment in the local economy as "payoffs" to minority leaders in order to gain support. Hmm. Let's back up one minute. Environmental justice proponents actively seek investment, economic gain and infrastructure development from large companies which choose to do business near minority communities. So Shintech and the state of Louisiana answered the call only to be met with screams of "foul." That was a revealing turn of events, but not as much as what happened next. Shintech managers finally caved, believing they would never satisfy the forces of environmental justice with their Romeville location. They did two things: they reduced the scope of their plans and moved the site to a more prosperous and more white town - Plaquemine, Louisiana.

"Foul" cried the environmental law clinic again. It was at that moment that Kuehn's law clinic finally and, in my mind, forever revealed its true anti-industrial, anti-capitalism bent. Bob Kuehn told local reporters "If Shintech thinks it can avoid the environmental justice issue by playing with the demographics, I think they are sadly mistaken." And, not to be outdone in resolve, Greenpeace issued a press release on behalf of local environmental groups saying the fight was "far from over," as they and other opponents vowed to follow Shintech wherever it moved.

Let's get this straight, the Tulane clinic and other ultras said Shintech was wrong to bring the cost of industry (pollution) to a poor minority community without also bringing the reward (jobs, money and taxes.) So Shintech offered jobs, investment and money. Tulane and others said no good. So Shintech took its pollution to a more environmentally equitable location (a more affluent, more white town). Tulane and others said no good to that too. The reality is clearly this: Big industry will never satisfy the ultras in the green movement, because the ultras are against big industry at all. They don't want clean industry, they appear to want no industry.

The governor and local industry realized the same conclusion and took the Tulane Law clinic to task, claiming the clinic was operating way outside of its intended scope. The clinic ought to help indigent individuals, not large communities, environmental organizations and Greenpeace. And the clinic ought to fight for individual rights (notably tenant rights, job rights, representation rights and the like) not fight against economic development. So the state sought and got a ruling from the a, which limited what the clinic could do in the environmental justice arena. The ruling states several things: for a community or group to get free legal help from the clinic, the majority (51%) of individual members of the community or group must be indigent as defined by income; law clinics may not in any case represent groups affiliated with national organizations; law clinics may not solicit cases or help residents organize into groups; and finally, law students may not appear before the state Legislature. At this writing the ruling is under fire, and some 200 law school professors have petitioned the court to reconsider. Bob Kuehn has resigned from the law clinic, but has stated the new rule is not the reason.

Moderates also didn't the threat from extremists to follow and fight Shintech wherever it went, regardless of efforts to work with local communities. And moderates were disturbed by the many other ultra extremist views making their way to the surface in the name of "justice." The belief that the earth is on the very brink of destruction and can be saved only by a total restructuring of society, for example, is not a moderate argument. So, when moderates heard arguments like that, the in-fighting began. No major industry, government body or status quo group wanted to hear that everything they had worked for had to be tossed out just to start over with a new set of rules. Effective negotiations for change don't begin that way. Disagreements between the moderate grass roots members and extremists erupted, sometimes publicly, as extremists gained press space and TV time. One result was that many moderate enviros removed themselves from the scene. Another result was the great body of potential new members, who were mostly moderate, thought twice about joining in. Extremism hurt recruitment, fund raising and membership of the movement. It was clear to me after several conversations with observers during Greenpeace's very visible Mississippi River campaign, that moderates found themselves trying to pick up the pieces after high profile extremist actions, which were usually followed by arrests of those involved. Before long, many moderates became disgusted and dropped out.

# Other Things That Hurt the Movement

Up to now, most things harmful to the movement I've mentioned fall more or less to motive - the underlying agendas which drove moderates away. However, some things are better classified as tactics which were applied across a range of agendas or motives. Some of these tactics were used and embraced by moderates - at first - because the motives seemed pure: a simple want for clean air and water. Once the underlying agendas driving the tactics became clear, many moderates felt misled and began to consider moving into some other line of volunteer work.

### Data Slinging

Although only a small number of environmentalists were extremists, one of their tactics creeped into the arsenal of moderates in a subtle way. I call it *data slinging*. By that I mean the act of throwing around scary numbers on deaths, disease and pollution in order to foment fear and provoke an angry response from grassroots recruits. Since the grassroots movement's strength derived from a largely lay membership, many members didn't have a clue about the existence of environmental data, much less how to interpret it or use it to further their cause. As a result, they were forced to listen to extremists who seemed to have command of the data and understood its meanings. At first, moderates, the media and politicians regarded the data as not only accurate but also as a valuable tool to get attention. Moderate and extreme enviros alike were able to shake things up with numbers that showed potential deaths and illnesses from high pollution rates, the inordinate length of time toxins remained in the environment once spilled, and how scientists could actually measure toxins in humans who had no significant exposure to them in their daily lives except for their air and drinking water.

Local and national extremists pointed out a list of wrongs seemingly supported by real data. In Louisiana, for example, enviros pointed out that cancer rates along the lower Mississippi River were out of line with national data, miscarriages in St. Gabriel were much too frequent, concentrations of neuroblastomas in children around Amelia were too highly concentrated to be random events, a chlorine leak from a single rail car could easily cross into the next state and kill hundreds of people for miles downwind, and we all lived dangerously close to deadly chemicals in nearby factories, rail cars and transport trucks. And other extremists played out identical scenarios in towns and cities across America.

The trouble was, a lot of this gloom and doom data didn't hold up. Even though extremists claimed loudly and repeatedly that cancer rates were too high and were caused by nearby chemical plants, epidemiologists - doctors who specialize in tracking illnesses - told me that they couldn't prove anyone got sick from pollution from a nearby plant. Dr.LuAnn White of Tulane University Environmental Medicine and Dr. Vivien Chen of Louisiana State University are two highly credible sources. They've studied pollution and cancer rates for years, trying to establish links where possible. They said we're all subjected to so many things in our lives in so many locations that tracking the cause of cancer to specific sources of air or water pollution is impossible. The epidemiologists told me they could measure how many people were sick in south Louisiana and find out how many of them lived near a chemical plant, but they couldn't rule out every other possible cause of illness, like cigarette smoke, auto exhaust, fatty and spicey diets, sunlight and fumes from dry cleaning, new carpet or gas pumps. They told me exhaustive studies were too difficult, too costly and would still not produce any solid answers. Plus, and this is the real travesty of cancer data slinging, their conclusive, nearly exhaustive efforts to understand potential industrial causes of cancer clearly showed that most cancer rates in Louisiana were normal compared to the rest of the country. The only cancer in which Louisiana had higher than normal rates was lung cancer, and the leading cause of that was smoking. According to researchers White, Chen and a host of other physicians and epidemiologists, the actual number one killer in Louisiana is heart disease, with lung cancer a distant second. Breast cancer is the next threat. No where near the top of the list are any diseases related to environmental pollution. In fact, in interview after interview with cancer specialists, heart specialists, and disease specialists I have been told few, if any, diseases which kill residents of Louisiana can be pinned on pollution, and most medical literature backs that up.

Well, extremists didn't let the facts stand in their way. Here are some specific examples of their most successful data slinging. Some time ago, business and industry nicknamed Louisiana's lower Mississippi River region between Baton Rouge and New Orleans the "Chemical Corridor," for the 80 or so sprawling petrochemical and chemical plants there which produced nearly 25 percent of the nation's chemicals. In the early 80s, during the BASF "lock-out" and union protests I wrote about earlier, union members and community extremists wanted to put pressure on the company to re-hire its workers, so organized labor began to shine a very public light on the company's poor environmental record. One union tactic was a citizens' march along the Mississippi River near BASF, carefully planned to pass in front of a newly printed billboard which read "Welcome to Cancer Alley." The news media covering the march jumped all over that, publishing and broadcasting photos of the march and billboard across the region and the country. That is how the name "Cancer Alley" started. No scientific evidence. No medical revelation. No data of any kind. Just a billboard. Greenpeace and local environmental leaders picked up on it and used the term in nearly everyone of their news conferences, press releases and interviews during the Greenpeace Mississippi River campaign in the summer of 1988, stating that here in South Louisiana, "We all live downstream." The intended appearance was that environmental leaders possessed actual data to dub the area "Cancer Alley," and the moderate membership didn't question it. Unfortunately the data consisted only of that single billboard.

The truest measure of success of that piece of data slinging was proven when President Clinton reassured a young boy on national TV that he knew all about "Cancer Alley" and, as President of the United States, he would handle it. In 1993 both President Clinton and Vice President Gore visited Louisiana and during separate appearances on national TV stated that Cancer Alley was a problem they would solve. Oops! In their efforts to let the country know how informed they were about Louisiana pollution and cancer, both spoke too soon, offering mini-lectures to national TV audiences based on false information. Their statements proved only that extremists had effectively portrayed the area to the media, and to the President, as a death trap.

When more reasonable and more accurate researchers at the National Centers for Disease Control, Tulane Medical Center and Louisiana State University Medical Center looked harder at the data (comprehensive examinations of all existing cancer studies, miscarriage rates, neuroblastomas and the like), they reached a very different conclusion. The researchers reported that, in general, residents of south Louisiana did not get cancer at a greater rate than anyone else. However, the researchers went several steps further to report that Louisiana residents who did contract cancer had a higher death rate than cancer victims in the rest of the country. The researchers concluded this was a result of poverty and poor education, not pollution. Folks who got cancer and couldn't afford early treatment usually waited too long to see a doctor, and had less chance to survive. But even more germane to the discussion of cancer's harm is that the researchers showed clearly and repeatedly that cancer was not the number one health threat in Louisiana. The real threat for both men and women was heart disease. Cancer ranked a distant second. And the form of cancer which was the most deadly was not breast cancer or neuroblastoma or pancreatic cancer of any of the cancers claimed to be environmentally related. The real killer cancer was lung cancer, and it was directly related to smoking. In fact, the single cancer which occurs at a greater than average rate in Louisiana is lung cancer in black males, which is directly related to the higher than average rate of smoking among black men in Louisiana. And guess what. Heart disease was also directly related to smoking, not pollution. In addition, a diet rich in fat and heavy spices was a contributor to obesity which was also directly linked to heart disease. Doctors and researchers have stated repeatedly the best and quickest way to improve health in Louisiana - that is to reduce heart disease and lung cancer - is to stop smoking, eat less fat and exercise. The environmental community has repeatedly ignored those very real facts. When Louisiana Governor Edwin Edwards reviewed the data, he wrote an open letter to President Clinton stating "there is no Cancer Alley in Louisiana." Edwards read the letter to a joint session of the Louisiana Legislature in March, 1993, which was broadcast across the state. It made no difference. Clinton never responded, and environmentalists never retracted the term.

Data slinging also occurred in the small town of St. Gabriel, La. Kay Gaudet, a pharmacist turned amateur pathologist, thought she noticed an unualuall high number of women coming into her St. Gabriel pharmacy with prescriptions for medication usually administered to females following miscarriages. Gaudet began counting and surveying her customers and other women in St. Gabriel and concluded the number of miscarriages in St. Gabriel was far too high. She got column after column of newsprint in local and national newspapers and large chunks of time on TV news programs after calling attention to her "data." The problem was, she was no researcher. When scientifically trained researchers who really knew how to count miscarriages (Tulane University pathologists) compared Gaudet's numbers with national data, St. Gabriel didn't appear to have a problem. Dr. LuAnn White, who managed the miscarriage project at Tulane, told me miscarriages were a rather common phenomenon nationwide, more common than people might think. Also, two important medical developments occurred about the time Gaudet started counting miscarriages. First, physicians learned to detect pregnancy only 3 to 4 weeks into the mother's term, which was dramatically sooner than the previous pregnancy detection of 8 to 12 weeks. Secondly, household pregnancy tests hit store shelves, allowing women to detect their own pregnancies. Before the early testing and home pregnancy testing, many women would become pregnant and miscarry during the first 8 to 12 weeks and not even know it. So, once pregnancy detection became easier, so did miscarriage detection. Kay Gaudet may have been unknowingly documenting the results of new technology, not the results of pollution. She was simply counting a normal level of miscarriages which looked high because many of them would have been previously undetectable.

About the same time of the St. Gabriel miscarriage stories, extremists also claimed they had data which revealed that scientists were finding a large number of toxins in human beings, including healthy individuals. Moderates were frightened when chemicals with names no one could pronounce turned up in blood and urine samples. It seemed as if we were all contaminated and didn't know it. More credible researchers, however, showed that human beings have always carried a number of toxins in their bodies from natural and man made sources, and the reason they were now showing up in greater numbers in human samples was because scientists had only recently learned to measure them in tiny quantities; 2 or 3 parts of toxin in a billion parts of water, for example.

Another case of data slinging hit rather close to home when, after reporting a story about groundwater contamination on my own news station, I learned I had been taken for a ride. First, a bit of background. Dow Chemical operates a huge facility near Plaquemine, Louisiana. The company makes all kinds of useful things - one of which gained them fame in Hollywood. When actor Bill Murray got "slimed" in the movie *Ghostbusters,* it was with a Dow Chemical product. Apparently, contrary to the picture painted by extremists, "Cancer Alley" did produce some completely non-toxic substances; Bill was literally dripping in it.

Although Dow Chemical sits in a rather rural looking setting, surrounded by large fields, it bumps up against the small towns of Addis and Plaquemine. Some long-time, family farms even sprawl right up to the company's fence. Dow is huge. It is actually a collection of several chemical plants each of which turns out products which are shipped to other chemical plants for use in their products.

Now back to more technical stuff. In Louisiana, the health department regularly tests water from wells for public water supplies but tests private well water only on request. A resident near the Dow site in Plaquemine, Louisiana requested such a test on his well, and the health department found traces of a toxic substance whose name I can hardly pronounce -
2-butanone - which is also known as methyl ethyl ketone (MEK). A local environmentalist of the more extreme

variety called my attention to the well test and the report of the chemical. I read the document she handed me, which appeared to be a legitimate copy of the test results, and reported at 6 pm on WBRZ-TV Eyewitness News that residents of Plaquemine were becoming concerned about their drinking water because one of their neighbor's wells had a chemical substance in it. Of course, the footage showed Dow Chemical of Louisiana looming in the background, and I reported that residents suspected the chemical came from there. So, trying to do my job, I had just reported a fear of contaminated drinking water to the majority of the news viewing population in the Baton Rouge metropolitan area. I had followed standard TV journalism procedures by reading the test results, quoting a local environmentalist about the well, quoting a DOW spokesperson and using compelling footage of homes and residents in the shadow of a giant chemical company. After all, the environmentalist who called my attention to the story was good at what she did. She usually knew what she was talking about and did her homework. When she said residents of a town just down the road from her own home were concerned about their water, I believed it. I also believed her statement about contaminants turning up in local wells and verified that with the report of the well she handed to me.

I am sorry to say that in my zeal to turn out a vigorous report, I didn't do enough homework. I took the data on the well offered to me and reported it. Actually, a testing laboratory later examined the well plus the surrounding soil and shallow groundwater. Their report revealed that indeed 2-butanone, along with a second chemical called methylene chloride and a third called tetrachloroethene, were found in soil samples and shallow ground water from near the well. The eye-opener, however, was the information my environmentalist did not give me.

**Goodbye Green**          **Duncan**

Routinely, when technicians test soil and water, they use "blanks" as part of their testing. These are clean soil and water samples which technicians carry to the site and handle exactly the same as the soil and water they are sampling. Test results from the blanks are compared to results from the actual samples. Technicians at the Plaquemine site measured both 2-butanone and methylene chloride in the blanks at the same levels as in the field samples. Both of these chemicals are solvents used for cleaning, glue making and other things and are found in typical laboratories. That means only one thing: the chemicals found in the soil and water samples were contaminants that creeped in during lab testing. This is a common occurrence; that's why technicians use "blanks" as a standard lab technique to spot it. The lab report I read described it as "typical laboratory contamination."

The third chemical in the soil, the tetrachloroethene, had a different origin. The lab report states it is associated with chlorinated hydrocarbons which, translated, means old, dirty, used motor oil. More on that later.

When technicians tested the well itself on the property - twice - they also found 2-butanone, but at concentrations above lab contamination. However, state Department of Environmental Quality records indicated no 2-butanone in any other well near Dow Chemical even though there were at least 3 groundwater monitoring wells nearby.

So here's the picture. It turns out that a rather sloppy truck maintenance and storage company used to operate on the lot behind the house with the questionable well, presumably at the knowledge or direct involvement of the homeowners and users of the well. Investigators found piles of old tires, 2 large steel storage tanks (capacity about 500 gallons), seven old storage drums filled with something which could have been used oil (technicians did not test the contents), numerous dilapidated vehicles (all of which presumably contained gas, oil and transmission fluid at one time) and "maintenance materials" associated with the buildings on the site.

Now back to the soil and shallow groundwater samples. Two chemicals in the soil and shallow ground water had been ruled out as laboratory contamination - the chemicals were solvents routinely found in laboratories. That leaves the third one - tetrachloroethene. Remember, it was associated with a chlorinated hydrocarbon. Hmmmm. Old storage tanks, a mystery oil of some kind in drums, old tires and years of spilled oil and fluids from trucks and cars. Also, it is known that mechanics used products which contained tetrachloroethene as degreasers to clean old motors. Gee, I wonder where the soil contamination came from? My environmentalist friend wasn't gracious enough to inform me of the cleaning and mechanical activities on the land near the well.

Okay, only one small problem left: the 2-butanone (MEK) in the single well on the home and truck repair company site. The MEK was definitely in a concentration higher than that caused by lab contamination The environmentalist accused Dow Chemical of the contamination and told me residents were worried about their drinking water. Well, Dow chemical didn't produce or use MEK, but technicians there knew where it could be found: in the glue used on PVC pipes; just like those used in the water well on the property. I wonder why my enviro friend didn't bring that up in conversation.

So, residents were worried about their drinking water (or so said my environmentalist friend) when no unusual levels of anything had been found in either the soil or the shallow groundwater, and only one well in the area had notable levels of PVC glue. Like bill Murray, my viewers and I had just been slimed, this time by data slinging.

When objective researchers began to acknowledge that these and other data simply didn't support extreme views, the general population, as well as existing and potential members of the grassroots movement, lost confidence in the movement's leaders and began to question their credibility. Many members dropped out, wanting instead to invest their time and money in more credible efforts and more reliable organizations.

To bring some sanity and rational thinking back into the discussion of the environment, Louisiana Governor Mike Foster's press secretary single-handedly stopped some veteran data slingers in their tracks. Activists from Greenpeace and the Calcasieu (pronounced kal' - ka - shoe) League for Environmental Action Now tried to grab some media attention and make some points by presenting press secretary Marsanne Golsby with a beautifully fried — but inedible — catfish. The enviros claimed they caught the fish in polluted Louisiana water, and that it was dangerous to eat, just like fish caught everyday by unsuspecting Louisianians. They presented the fish to illustrate how the Louisiana hunting and fishing life-style was on the decline. Well, knowing exactly what was in store for her when a Louisiana native prepared a catfish (Louisiana isn't known for its food for nothin' ya'll) Golsby did what came naturally. She pulled a plastic fork from her pocket and ate the fish for lunch - in front of a stunned group of environmentalists who begged her to stop as they watched their own data get slung (or, rather, eaten.)

**Spreading the Blame**

Up to this point, I've placed most of the blame for the death of the grassroots environmental movement on extremists within the movement and on the encroachment of pseudo-environmental organizations (trade unions, civil rights groups, conservation groups) which started a fight for control of the movement's agenda, membership and dollars. However, at least one moderate grassroots strategy also slowed the progress of the movement: the effort to spread the immediate blame for pollution to everyone. What I mean is this. Rather than point to specific air and water pollution sources - which inevitably were big industrial sites - conservative environmentalists believed it was more politically correct to accept blame and get the rest of us to accept it too. There was merit to that argument. If you added up all the paintbrushes we hosed off in our yards, all the engine oil we spilled, all the chemicals we used to kill mosquitoes and all the plastic we threw on the ground, we actually did contribute a lot to pollution. However, the facts seemed clear to me as a reporter and clear to me now, the quickest way to clean air and water is to stem the flow from industry first and work on our own trash second. Unfortunately, the strategy of spreading the blame didn't prioritize pollution; it made all levels of pollution equal (i.e. dropping litter was a harm equal to the industrial belching of toxins into the air.) This strategy contributed to the movement's slow down and death because it eliminated clear targets and goals.

The idea for spreading the blame emerged sometime around the revival Earth Day on April 22, 1990, when moderates came to believe that the best way to empower the movement was to increase every American's interest in it. The result was an enormous effort to teach us that we were all part of the pollution problem and should, therefore, be part of the solution. That's when we all learned to recycle and asked our community leaders for recycling plans, curbside bins and regular household pick-up of recyclables, complete with new taxes to pay for it all. You might remember that this effort was so successful it caught industry off-guard. Few companies had the necessary equipment, expertise or collection systems to actually use growing mounds of recyclables. Cities found themselves collecting plastic jugs, newspapers and glass in huge quantities merely to stockpile it, or worse, send it to the dump without telling anyone. State and city planning commissions also voted to start ozone reduction measures which affected everyone, not just the big polluters. Car pool lanes, bus systems, restrictions on back yard barbecues, more rules on keeping car exhaust clean and more stringent inspections to monitor cars became facts of life. We saw public service announcements, flyers in our electric bills and inserts in the newspapers on how to recycle our tires, engine oil, car batteries and refrigerators. Shade tree mechanics and do-it-yourselfers were stymied when it came time to repair or recharge their car's air conditioner. New rules to protect the ozone layer pulled freon off the shelves and put it only in the hands of certified repair shops equipped with expensive freon capturing and recycling equipment.

As a society, we decided all of this was necessary, and we agreed to pay the price for it. Gas prices rose to cover the costs of new fume capturing devices on the pump. Auto dealers and parts stores had to add a dollar or two to the price of a new tire or car battery to pay for recycling the old. We also paid greater taxes for community ozone reduction measures, and we watched the prices of our newspapers and other goods rise to cover the new costs of recycled paper and packaging. And in an bit of irony, we all paid taxes to start curbside recycling plans, even though the bottles, cans and newspapers we collected at the curb were a resalable commodity. Apparently their market price didn't pay for collecting, sorting and transporting them, or so we were told, but we recycled anyway and were glad to do our part.

Recycling and saving the planet were great, easy-to-understand goals which played well in classrooms, so school children also began putting on a record number of plays, songs and skits about saving the earth, and their literature began to carry environmental themes. But spreading the blame for pollution to our school children seemed a little much. The message of one video I pulled from my children's elementary school library collection was crystal clear. First, it listed a variety of pollution concerns: solid wastes, millions of tons of discarded paper, glass, bottles, plastic containers and aluminum cans, overflowing landfills, spent natural resources, wasted energy and air and water pollution. That pretty much covers the gamut, I'd say, except for specifically naming a couple of Superfund sites. The tape then went on to teach: "Each of us shares in the responsibility for creating these problems, and each of us suffers from their consequences. But we can all help in the search for solutions." (*Reducing, reusing, and recycling: Environmental Concerns*, Rainbow Educational Video). On the face of it, no one can argue with the general message of cleaning up the earth, but I see two problems with putting the blame for such diverse pollution problems - and the solutions - on our children. First, I don't think we are all responsible for the majority of water and air pollution. I don't know about you, but my children and I don't send millions of tons of toxins up smoke stacks or into the Mississippi River or down into injection wells. Spreading the blame for all those things diluted the message and blurred the goals of the grassroots movement. We chose the easy, seemingly obvious "fixes" without real thought. For example, many new parents (my wife and I included) chose cloth diapers over disposables because the environmentalists at the time said the thousands of diapers we threw out were creating a mountain, literally, of untreated sewage. Diaper services sprung up across the nation - something our parents were familiar with long ago. However, thoughtful researchers who evaluated both disposables and washables came to an interesting conclusion. The cost of washing diapers (with hot water), disposing of wash water (complete with detergents), drying diapers in gas or electric dryers and delivering them (in trucks) impacted the planet sufficiently to make their choice as an earth saver less obvious. Once that word got out, diaper services seemed to dwindle. My yellow pages no longer lists any. Second, videos like the one I pulled and countless other tapes, books, songs and materials which teach our children that they were responsible for cleaning up pollution created somewhat of an "us against them" mindset. Children were taught that they - the good guys - would have to grow up to confront big business and industry - the bad guys - and somehow win the environmental war. That encouragement of confrontation without sound reason is a tactic that chased moderates out of the movement.

The moderate environmental message that "everyone is the cause so everyone is the cure" propelled the grassroots movement onto the front pages, but at the same time took a lot of heat off big industry. Suddenly, environmentalists were claiming every one of us was the cause of dirty air, land and water, which made it harder for us to gather clout and momentum to focus on those huge smokestacks and demand that the industrial operators clean them up. Industry execs simply retorted, "not so fast, we're not the only cause of pollution here." The fights in city halls, legislatures and government agencies shifted from grassroots grannies pointing their fingers at big industries to industries pointing back, and everyone else pointing at each other. Rather than maintaining the forward momentum of the movement to clean up specific hazardous sites, target individual companies or improve industrial practices, this overemphasis on individual, everyday citizens as the cause of pollution may have been a contributor to the falling off of enthusiasm, involvement and membership in the movement. Experts in public relations or group organizations agree that when goals are blurry, motivation and results are almost impossible, and I argue that few people want to join a movement just to learn they are the cause of the problem. Once we all became the cause, the movement's goals became blurry and momentum waned.

Finally, a large dose of extreme exaggeration distorted the initially moderate messages of "everyone's responsible" and "children are the good guys" into extreme, almost angry messages - a turn off to the moderates who promoted them. A good example of this is the extreme "save the earth" environmental message crafted for children by a variety of organizations, authors and individuals. Faced with an onslaught of stories, plays, lessons, handouts and videos featuring big, ugly smoke stacks, crowded highways and dirty manufacturing, any normal, healthy 5th grader would be embarrassed to admit his mom or dad worked for Dow or Exxon or GM.

The reality is that the "save the earth" theme of much of the elementary literature I have seen has been played up at the expense of the truth. Children may be taught recycling is good and industrial pollution is bad, but fail to learn their tennis shoes, book bags, notebooks, desks, pencils and paper are all industrial products, not to mention the medicines they are given to get well when they get sick. Children are actually learning that because industry causes pollution, industry is bad. That simplified - and wrongheaded - message was carried to an unbelievable extreme by Allan Schnaiberg and Kenneth Alan Gould in their book *Environment and Society, the Enduring Conflict*. In the book, they actually suggest we all cause the pollution problem when we strive for better jobs with higher pay (the profit motive, you know, leads to all the earth's ills.) So we should apply for less credit, opt for jobs with less pay or go work for the government because all of those ideas put less demand on private sector to produce goods. Instead of the extreme notion that industry is inherently bad, the message ought to be industry which enhances our lives is inherently good, but some bad practices ought to stop.

However, I think the "industry is bad" message is winning and becoming ingrained in school literature and the minds of our youth. Take note of the recent popularity of the Susan Jeffers book *Brother Eagle, Sister Sky*. It is an adaptation of a long-ago speech reportedly given by Pacific Northwest Chief Seattle when Governor Isaac Stevens arrived in Washington during his search for a rail route to the Pacific. That was about 1853 or 1854. On its face, the book is a beautifully illustrated story of the demise of America from the perspective of native Americans. It is written in poem or song form as an ancient native American story and contains passages like, "When the view of the ripe hills is blotted by talking wires..."

Jeffers' version paints a crystal clear picture: the white man ruined the land, polluted it and overdeveloped it, and native Americans were the only ones to recognize it and feel the loss. The problem is, Jeffers' version is a rewritten version of a rewritten version of the original; a third generation, if you will. It appears Jeffers may simply have based her book on a version of the speech displayed at the 1974 Spokane Exposition. Jeffers called it an "adaptation," stating in the preface,

> "What matters most is that Chief Seattle's words inspired - and continue to inspire - a most compelling truth: In our zeal to build and possess, we may lose all that we have." - Susan Jeffers

However, the anti-pollution, anti-industrial message was not even present in Seattle's original speech, according to critic Jon C. Stott in his publication *Native Americans in Children's Literature* (Onyx Press, 1995). According to Stott and other historians, because Chief Seattle was forced to cede lands to the government for a railroad right-of-way, his speech was therefore an embittered statement against the usurping of land, not about the environmental degradation of it.

Stott gives a thorough and convincing argument that Jeffers took too many liberties with both the illustrations and the factual content. The Pacific Northwest chief is illustrated as a central plains leader, for example, and Jeffers' loose writing matches the inaccurate illustration. Most North American natives considered the earth as mother and the sky as father not sister as in *Brother Eagle, Sister Sky*. Jeffers' history may be a bit off as well. According to my encyclopedia, the telegraph was patented by Samuel Morse in 1844, and the word "telegram" wasn't coined until 1852. Although news accounts report the telegraph had reached every state east of the Mississippi by 1848, they probably hadn't yet reached Washington state by the time of Seattle's speech. Remember, Governor Isaac was there searching for suitable passage for a railroad to the Pacific. Telegraphs were operated by rail companies and were placed along railroad tracks. So, according to Jeffers, Seattle was likely bemoaning the "view of the ripe hills blotted out by talking wires," before the first railroad track or first telegraph wire even showed up! Jeffers adapted an adaptation, a process which misrepresented the theme, message and intent of the speech as well as the physical appearance of the peoples and the land, just to get us all to realize a "truth" that industrialization, American style, is bad. Remember, kids all over America are reading this stuff, and I bet when teachers assign it, they don't assign Stott's critique of it. When I asked some high-schoolers at the school library if they had read the book they said "yes." When I asked if they were aware of the controversial approach Jeffers took, they had no idea.

At the same school, I had a chance to interview a top student who had won awards for her past environmental projects and was already knee deep in some more. I was surprised to learn that her belief and her self-appointed mission was clearly anti-industry, yet she had never had a chance to meet or talk to anyone operating or managing any of the dozens of local chemical companies. If the students I spoke to then and the students I meet regularly are any indication, and they are, the slanted "industry is bad" message is winning the hearts and minds of our youth. That,

I believe, is the worst result of data slinging and is also what drove many moderates away from the movement.

**Superfund**

The idea of spreading out both the blame and cost for cleaning up pollution also caused what I believe was one of the largest boondoggles ever for litigation lawyers - the "Superfund." The official name was, of course, less dramatic. The Comprehensive Environmental Response, Compensation and Liability Act of 1980, which is fondly referred to by attorneys as CERCLA (pronounced SIR - kluh), created a multi-billion dollar fund, set aside by your government and mine, to clean up the most toxic waste sites in the country. That promise of clean up required communities to specifically identify toxic sites and submit detailed applications to get them included on the federal Superfund list of designated sites. Typical sites included old oil storage yards, abandoned city dumps and old industrial sites. The application process, and presumably the reality of a finite amount of money available for clean up, placed enormous pressure on communities across the nation to act quickly and created a high stakes competition for site acceptance and superfund money. Investigators, attorneys and clerks suddenly found lots of work in the business of identifying potential superfund sites and in submitting thousands of applications for superfund money. Part of CERCLA itself even contributed to the avalanche of applications for clean up, because the act actually made money available to environmental groups so they could hire specialists to investigate and evaluate potential sites for superfund listing. That meant the fund actually paid money to create cases on which to spend more money. Theoretically, once the Environmental Protection Agency approved a site for Superfund listing, millions of dollars would pour in to clean it up, but the fund shelled out billions of dollars and did virtually nothing, except to pay for a boat load of EPA attorneys, private attorneys, specialists, technicians and investigators to examine every application for superfund money. The money spent was so enormous and progress at toxic sites so scant that by 1997, critics, citizens and town leaders were on Capitol Hill pushing congress to rewrite CERCLA to ensure money was actually spent to clean sites up. But the mess got worse.

The Superfund law had a neat little section which EPA and private attorneys used to clog and foul federal and local courts rather than to unclog and defile the nation's waste sites. The law allowed EPA to use Superfund money to research potential sites and allowed individual state environmental agencies to do the same. That "research" was more than just the technical study of which wastes were present, how much was there and how to clean the site up. EPA could also use Superfund money to dig through courthouse records, subpoena witnesses and to seize documents and the like to determine the names and whereabouts of anyone - company or individual - who ever used the waste site, operated the waste site or in any way had waste deposited in the site. Once discovered, the EPA called these found individuals and companies "PRPs" for Potentially Responsible Parties. The EPA could then order the PRPs to clean up the site at their own expense. If PRPs ignored the order or refused, the EPA could use the Justice Department to sue, forcing PRPs to pay for the clean up and court costs, thus saving precious Superfund money for sites where the EPA could find no PRPs. That wasn't a bad gig in the eyes of the EPA, so it found itself spending more money on research and attorneys to write orders or file lawsuits than to clean up sites.

Now here is the kicker. Finding PRPs was hard, time consuming work, especially if the hazardous waste site was 50 or 60 years old and had been out of operation for decades.

Identifying the operators of a dilapidated oil storage facility, abandoned factory or overgrown commercial landfill and then identifying all the individuals and companies who may have sent oil, supplies, waste or garbage to those places took time and money. Well, the writers of CERCLA must have been well coached, because they gave the EPA a way to bail out of the work. The law included a section which stated if the EPA could identify one PRP who willingly followed the order to clean up the site, then that PRP, whether a company, individual or even government agency, was allowed to sue any other PRP to recover not just the clean-up costs, but three times the costs. That's right. A cooperative PRP could sue anyone else it could identify for three times the cost of the actual clean-up and court costs. Attorneys call that triple damages. The PRP, therefore, could literally invest in a clean-up operation, triple its money, pay off the private attorneys' pockets who found and sued the other PRPs and walk away with a profit on the whole operation. In the words of one of my own state's officials, "There's big money to be made in this thing."

So to prevent a whole lot of legwork and to save precious Superfund money, the EPA began to identify only the largest PRPs and let them finger the smaller guys. An entire generation of attorneys could make a living and send their children to law school for the money spent on the mountains of litigation. Not surprisingly, following the lead of the federal law, state environmental agencies did the same. States had their own waste site clean-up money and their own procedures, but because they were subject to federal law, states pretty much mirrored the CERCLA regulations. Typically, if a state felt a certain site was simply too large, too complicated or too expensive to clean up itself, it could apply to get the site on the federal Superfund list and tap all that federal money. The EPA then just cranked up its legal proceedings to identify responsible parties and issue orders, maybe even duplicating the state's efforts and therefore wasting more money.

So rather than create a fund which would simply clean up superfund sites, Uncle Sam created a fund that paid for environmentalists and EPA attorneys to sue companies to clean up who could turn right around and sue others, who could then turn around and sue others, and on and on. Your government and mine created and paid for a system whereby anyone accused of pollution at a superfund site was encouraged to go out and sue anyone else they could think of. And here's the shocker. Any person or company who sent waste to a dump could wind up being sued to clean up that dump even though they may have followed every environmental law in place while the dump was operating.

Because the vast majority of superfund sites were old and abandoned private and municipal dumps, some of which were 50 years old, they contained thousands of tons of all types of waste deposited there under the sparse laws in place during their operation. Those few laws called for little if any identification of the waste and called for few technical specifications of the landfills themselves, like water proof linings, or groundwater protection systems or adequate cover and fill. Therefore, tons of unmonitored, unidentified hazardous wastes lay buried for years in community landfills across the country under relatively thin layers of dirt and sod. Many began to leak. The old dumps simply were not designed to contain toxins for very long. They had no thick plastic linings, no clay or concrete cap to keep rain water from seeping in and no pumps and filters to pull out rain water that did leak in. So, even though many of these dumps were built properly under laws in existence in the 40s, 50s, 60s and 70s, and many of the contributors to them were following the regulations in existence at the time, the new CERCLA was still able to force individuals and companies to clean them up under new environmental laws. This retroactive law seemed patently unfair to many caught in its long arm, but it was the law nonetheless. Fortunately for corporations identified as contributors, they had the EPA's blessing - backed up by law - to recover their costs and a lot more by dragging in others, and they took full advantage of it.

So picture this: state and EPA officials or their attorneys combing through old files or records of closed dumps; field agents out raking, searching and digging in the abandoned dumps for boxes, drums, labels, cargo slips, or anything else which would determine who sent stuff there; clerks sending out a wave of notifications to the largest companies identified as potentially responsible parties with a demand to clean it up; then those companies, sometimes faced with billions of dollars in estimated clean-up costs, doing exactly the same thing to identify others they could sue to spread out clean-up costs or even make money by collecting triple damages. That was exactly how numerous individuals, small businesses, a local café owner and an old folks home landed in court with a huge bill for the clean up of a toxic waste site. It seems that each of them had sent some trash or garbage to their local dumps, and those dumps eventually ended up on the Superfund list as hazardous waste sites. Using a return address from perhaps a single discarded envelope found amidst the rubbish, leaking drums and oily soil of a dump site, or using a signature on a waste hauler's receipt, the EPA could mail a lawsuit and a bill for the clean up right to the mailboxes of grassroots Americans. Thus, unaware citizens were fingered as villains and wound up as defendants in EPA lawsuits to clean up a multi-million dollar mess. It sounds ridiculous I know, but keep in mind that in 1997, nearly a quarter of the 1200 listed Superfund sites were simply old, abandoned city dumps. All kinds of trash and toxic substances ended up in them from all kinds of people and businesses, and it seems all those people were being dragged to court to explain their evil deeds.

Wall Street Journal reporter John Fialka negotiated the quagmire called Superfund to reveal the frustration of everyday people caught in the government trap. One October day, in 1995, Barbara Williams went to retrieve her mail as part of an otherwise normal day managing her small café, the Sunny Ray Restaurant, outside Gettysburg. In her mail, she found some very unexpected documents addressed to her from the U.S. Department of Justice. Barbara Williams was being sued for sending toxic industrial wastes to the local dump even though what she had sent was probably no more toxic than food scraps. Apparently the Department of Justice had been working overtime. Williams learned she was only one of more than 800 people hit by a shotgun blast of lawsuits coming from the barrel of an even larger lawsuit dubbed U.S. vs. Keystone, filed under the Superfund law. Naturally, the unsuspecting residents, small business owners and others spent an awful lot of time just figuring out what was happening to them.

They learned that years earlier an old city dump named the Keystone Sanitary Landfill had been closed but had begun leaking poisonous heavy metals and other toxins into the groundwater. Apparently, residents and legislators in Gettysburg had fought to get Keystone on the Superfund clean-up list mistakenly thinking the landfill would actually get cleaned up. Keystone was accepted by the EPA and placed on the national Superfund list. No one knew how all that stuff got in the landfill or who put it there, but no matter. The Superfund law let the EPA sue whomever it could identify as a user of the landfill, so that's exactly what the EPA did. The agency set its sights on the 13 largest businesses it could identify and then hired the Justice Department to go after them. Voila! The United States vs. Keystone. Those 13 businesses took the first hit when U.S. attorneys filed the estimated costs to clean Keystone up and a demand to do so. You can imagine the reaction.

"Whoa!" said the 13, "We're not the only ones."

"Okay," said the EPA, "Then tell us who the others are."

The 13 then hired a bunch of lawyers who found 120 more businesses to sue. Like a spreading virus, those 120 middle-sized businesses then found hundreds of smaller businesses who could sue even smaller businesses. That's when a local bowling alley, a motel, a church run retirement home, and Barbara Williams all found themselves in court. You can almost imagine those owners' looks of confusion, then fright after they were handed the bill to clean up a Superfund site which had been identified by the U.S. Government as an immediate hazard. SunnyRay owner Williams' share of the clean-up was $75,000, but she learned the PRPs (potentially responsible parties) who sued her might settle for $25,000.

Meanwhile, the Philadelphia lawyers hired by the Justice Department and the first level of PRPs are, at the time of this writing, trying to manage more than 800 separate cases. The U.S. vs. Keystone has become a huge milk cow for Philadelphia lawyers who are prosecuting the case. According to reporter Fialka's sources, some estimates reveal the attorneys alone will soak up $30 million on legal actions. That means of course, $30 million less to clean-up the dump. I argue that the EPA might have dug the dump up, incinerated the whole mess and then put back the clean ash and dirt and sodded the whole site for that kind of dough. Instead, the EPA chose to hire the Justice Department which in turn hired Philadelphia lawyers to identify the largest PRPs who hired lawyers to chase down café owners and retirement homes. But that was just in Gettysburg. The EPA estimates that 25,000 small businesses and individuals have faced action brought against them under the Superfund law. Yet, as of about 1997 the EPA has cleaned up only about 125 of the 1300 or so sites on its list. And, according to the General Accounting Office, the length of time to actually get a Superfund site cleaned up in 1986 was about 2.3 years. By 1996, that length of time had stretched to 10.5 years. The wheels of justice aren't turning slowly, they are bogged down all together.

Is it any wonder the American people have lost faith in the grassroots environmental movement? Environmental activists nationwide fought for the Superfund, or something like it. Now it's being used against the moderate, grassroots grannies who were told it would clean up their country. But there's more. The Superfund actually pays for local environmental groups to hire specialists to evaluate their own pet sites in order to determine whether they also should be listed on the Superfund. So who are these specialists? They are usually semi-pro activists who have taken it upon themselves to become local experts who can woo grassroots enviros to hire them. So it seems the Superfund is lining the pockets of lawyers and environmental activists at the expense of the vast, moderate membership of the movement. It's no shock that disgusted, middle-class and moderate members left the movement to return to their children's soccer practice, their hobbies or other interests. They had seen enough. They watched their local superfund sites continue to stagnate while hundreds of their own kind were sucked into courts to pay up or fight. Eventually, even the radical enviros became disgusted. Conversation in Washington has since turned to the possibility of re-writing CERCLA or scrapping the Superfund as an almost completely ineffective means of cleaning up toxic waste sites.

I've come to the conclusion that the progress of any grassroots movement lies in a long, long series of small victories. At first, getting site after site accepted on the Superfund list was just that, a series of small victories. But the swamp beast called "Superfund" took all those hard fought battles and turned them into one giant loss, because the sites lay dormant for years no matter now many times they were picketed. Leave it to the slow wheels of government to have wrested the hope of action from the hearts of thousand of Americans. And when Americans lost hope, they left the movement.

**Agents Provocateur**

Let me take a moment to say something here before I move on. As a reporter, I am much more comfortable writing fact than speculation or rumor, but the shadowy notions of provocation and front groups don't lend themselves to clean, neat corroboration. And many a story began as a hunch or suspicion only to be backed up by facts revealed later. Parts of this one are the same. Some of it is left to alternate explanations, but until I hear them clearly, I'm sticking with my conclusions.

My seventh grade English teacher taught me to never write in a language I didn't know or which my readers couldn't read. But I can't help it now. The French term for people who stir up protest in order to confuse and entrap - agents provocateur - seems entirely appropriate. These were the people put in place to look like legitimate environmental activists, but who were really agents of big business. They were a blow to the grassroots movement primarily because the media didn't adequately report their standing - or lack thereof - in the environmental community. They blurred the agenda, confused the issues and lessened the impact of legitimate organizations. Worse, in some cases local legitimate organizations may have tolerated the front groups, an act which led to further dilution. It stands to reason that if a legitimate group, whether environmentalists or some other interest, allows impostors to run around appearing to take up the fight in order to advance a secret agenda, then the original group's legitimacy and credibility would be hurt if anyone found out, which is exactly what happened. Local environmentalists' failure to blow the whistle on impostors did hurt the movement.

A good example of front group use occurred during the height of the environmental movement in Louisiana. Two major hazardous waste incinerators there were fighting two battles; one against environmentalists to get the permits they needed and one against each other for customers. It seemed that there was only so much toxic goo to be burned off, and each company needed to take a bite out of the other's business. They competed heavily for large industrial contracts to dispose of everything from oily sludge, to out-of-date pesticide to contaminated soil. The two incinerators were Marine Shale Processors in Amelia, Louisiana and Rollins Environmental Services, located in Baton Rouge, a couple of hours up the road from Marine Shale. During the movement, both were under enormous pressure from local, state and national environmentalists to stop incinerating waste and sending poisons into the air. Both were also trying to get state and federal permits to stay in business, which gave protesters more opportunity to shut them down through political means as well as through public actions. Each company reacted quite differently.

One big difference in approach by the two companies was that Rollins appeared to play by the rules. It began to handle the press and TV crews with candor, to attend community meetings about its operations and to follow the rules and regulations governing hazardous waste incinerators while applying for a permanent operating permit. (During the events described here, Rollins was operating under a temporary permit.) Marine Shale, on the other hand, seemed to operate by its own rules. It kept its distance from reporters, shouted back at protestors and claimed it was not even a hazardous waste incinerator, but rather a hazardous waste recycler. That last point was extremely important to Marine Shale, because recyclers could operate under different, less stringent laws than incinerators.

Here's a quick, technical comparison. In Baton Rouge, Rollins used a large, rotating steel tube to incinerate waste.

The tube was not quite horizontal, with one end slightly higher than the other. Conveyor belts carried barrels of hazardous waste up and into the higher end of the rotating tube. As the waste tumbled and slid slowly down the inside of the tube to the lower end, gas flame jets literally incinerated the stuff at about 2000 degrees. Theoretically, the incinerator vaporized the waste, cleaned the vapors with a water spray coupled with huge filters and then sent the clean exhaust up a stack as simple carbon dioxide and steam. The only thing that came out of the low end of the rotating tube was some ash. Rollins inspected and tested the ash before burying it in a landfill.

Marine Shale did exactly the same thing with a much bigger, double barreled incinerator, but claimed it didn't have to meet the same requirements as Rollins or undergo the same rigorous inspections. The company's reasoning lay in it's claim that the solid residue from its rotating incinerator ash was inherently different than that of Rollins. While Rollins called it "ash" and buried it in a landfill, Marine Shale called it "aggregate" and tried to sell it as road building material. Indeed, Marine Shale's ash looked like black, ceramic chips and from a distance resembled mounds of black gravel, but to my knowledge the company never sold a pound. The Louisiana Department of Environmental Quality never approved it as a building material. The aggregate simply accumulated in huge piles which grew bigger between each reporting trip I made to the company. Undaunted, however, Marine Shale claimed its aggregate was indeed a suitable road-building material, and because the aggregate had come from hazardous waste, the company was therefore a recycler not an incinerator. So, while Rollins, as a legally operating incinerator, had to meet a score of air, water and land pollution requirements by undergoing rigorous inspections and tests designed for hazardous waste incinerators, Marine Shale claimed it could operate under more lax rules designed to encourage recycling. (The rules were real, by the way, but were meant to apply only to small incinerators that a chemical plant might operate to fire its own boiler to produce steam and electricity.)

Environmentalists didn't buy the recycling argument of Marine Shale and neither did the Environmental Protection Agency which hit Marine Shale with about a two million dollar fine. Greenpeace and other groups were quick to praise the fine, pointing to Marine Shale as a sham recycler. The protestors claimed Marine Shale management had been simply trying underhanded tricks to circumvent the law and had finally been caught. But the recycling argument, apparently, was not the only trick up Marine Shale's sleeve.

Let's get back to the intrigue. In the 80's and early 90's, both Marine Shale and Rollins were facing very close scrutiny from the public. The state environmental agency would grant permanent operating permits only after lengthy public hearings, and Louisiana residents took advantage of every opportunity to pack local public hearing rooms to call both companies polluters and call Marine Shale a sham recycler. The problem for both companies became how to ease the pressure at the public hearings by minimizing public attacks and gaining some credibility.

Rollins sent a whole new management team to Baton Rouge with the claim they were going to operate in a more open manner. The media, which was normally stopped at the front gate, would be allowed in. In fact, the company once took a huge risk by allowing me to view and report on their brand new shredder. This machine contained a nest of rotating, overlapping saw blades. Metal barrels of flammable liquid waste were supposed to drop into the top where the blades would chew them up and spit out two things: the liquid in the barrels, which was sucked into the incinerator's feed lines to become fuel for the gas jets, and the little metal shreds, which went to the rotating tube to be vaporized. But as luck would have it, the shredder didn't work when workers demonstrated it to me. It was actually a pretty funny sight. Managers in coats and ties struggling along with the hard-hatters to get this thing to shred cans. It wasn't really a big deal, a blown fuse as I recall, and I knew any new device was subject to a problem or two so the snag never made it into my story. Besides, management assured me the shredder wasn't going into operation until it was working properly, and I had no reason to doubt them. The point is that few companies, even forthright, clean, "good" companies were willing to risk a media inspection of a new device until it was up and running perfectly. Rollins' effort actually turned into pretty good public relations. They showed a little humanity, warts and all, or so it seemed.

When reporters were not looking, a concerned citizens' group rose up to oppose Marine Shale in public hearings and in the press. That group called itself the Hazardous Waste Treatment Council and pressed the same arguments as environmentalists had; Marine Shale was an incinerator and not a recycler, the company needed to clean up the piles of aggregate, stop spilling toxic runoff into a nearby bayou and start following all the rules which applied to commercial hazardous waste incinerators. Well guess who helped fund and organize this so-called environmental group; Rollins and other hazardous waste treatment companies. Good old corporate competition had taken a whole new approach. I guess corporations were catching on to that grassroots thing, too. Talk in boardrooms and executive offices had begun to include the ins and outs of industry forming and supporting its own grassroots support groups. Although I don't know of any corporations who announced their involvement in such groups, some provided a straight forward answer when asked about their activity. They said since jobs were on the line, their employees should be as well represented as the environmentalists on the picket lines and in public hearings. Rollins management obviously felt the same way, but they jumped in with the Hazardous Waste Treatment Council in a slightly different way. The group didn't exist to support Rollins in the permit process, it existed to oppose Marine Shale in its efforts to get operating permits.

Not to be outdone, Marine Shale apparently also had a trick or two up its sleeve but took a drastically different approach to public relations in general. Unlike Rollins, Marine Shale never tried to polish its image in the news media or attempt to be good guys or reach out to reporters with demonstrations or tours. In fact, company management remained absolutely belligerent when dealing with the news media. For example, company president Jack Kent once announced in a television news story that if Greenpeace showed up to try to shut the company down, he and his employees would "be there with ax handles painted green, so they could see them coming." And that's pretty much what happened. Several weeks later, Greenpeace activists in rubber boats armed with water pumps approached Marine Shale from an adjacent bayou to suck polluted bayou water up and spray it back onto Marine Shale property. Sure enough, tough looking Marine Shale hardhatters met the boats at water's edge and beat the Greenpeace leader bloody. Television stations in Louisiana had it on the 6 o'clock news, reducing further Marine Shale's already tarnished image. Apparently, Marine Shale didn't care about projecting a good front to the news media. However, Jack Kent did appeal directly to the public with a "good guy" ad campaign. It was rather strange to see such a violation of generally accepted public relations strategy which usually calls for a consistent message. While a red faced Jack Kent was on the news threatening Greenpeace with ax handles, a warm and inviting Kent was in the commercials claiming his company was not a polluter but instead "cleaned the stuff up." I suppose you could call that a two-pronged strategy, but what happened next was apparently a third, more sinister prong.

I believe Marine Shale was the likely force behind covert actions in Baton Rouge. The company may have backed -

and perhaps even created - a citizens group whose sole purpose was to turn the heat up on Rollins during its permit application process. Here are some key observations. First, a new environmental group, called Christians for Good Government, appeared on the scene right in the middle of the Marine Shale/Rollins fight for permits and business. The group's main strategy was to picket Rollins, informing the local media before hand of course, and use every opportunity it could to get some press space and air time to complain about Rollins. That wasn't suspicious in the least, in fact that was a very usual way for local environmental groups to act. But what got my and other reporters' attention was the fact that the group was tiny - perhaps a dozen or so people - and suddenly found a high powered attorney to represent them, the former district attorney of Baton Rouge. Next, they also rounded up a public relations firm in Baton Rouge to create, manage, advise and represent them. A young man from the public relations firm organized the group to picket Rollins, informed the media, met reporters on scene to coordinate interviews, and even helped the pickets get transportation.

Now let me say this. Marine Shale's connection to Christians for a Clean Environment is not revealed in any documents that I can get my hands on, nor would they be. In fact, the Christians for Good Government's attorney denies it. However, the circumstantial evidence is rather convincing. Just when Rollins was applying for a permanent operating permit, Christians for Good Government appeared. I first learned of them through a phone call from the young public relations man who "tipped me off" to a protest. When I arrived at the gates of Rollins, a group of about 8 to 10 people were there, mostly elderly black men and women. They told me they lived in the nearby Alsen community, which can only be described as a predominantly poor, black community. However, one young, stylishly dressed man in their midst was pushing an elderly man in a wheel chair. The young man just didn't seem to fit in, so I approached him. When pressed, he told me he was with a public relations company and his job was simply to help this group with transportation and a few other details. The glaring discrepancy was that he drove a small sports car. I didn't see how it could be much help with transportation, and, in fact, I never saw him drive anyone to or from any rallies. Hmmm.... a young man, driving a gleaming sports car and employed by a public relations agency shows up to "help out" a small group of elders from a far off community; something didn't seen quite right with that picture. Plus, he had worked the media, including me, to get coverage. It was the first local environmental group I knew of which had a professional public relations firm helping them battle a corporation. I'm ashamed to admit I didn't pursue those little nuggets as hard as I should have at the time, but I did confirm the young man worked for a public relations company in Baton Rouge and made a mental note to check it out further later.

Shortly after that protest, the former district attorney of Baton Rouge, Ossie Brown, appeared in the news as the group's spokesperson. Okay, you make the call. Did this little group of poor people hire both a public relations firm and a high-powered attorney? Could they afford to? Nope. But Marine Shale could, and they had the motive and willingness to strike out at their competition by bullying, beating and otherwise acting like anything other than a model corporate citizen. I am not the only one to believe that Christians for Good Government was getting funded by someone, and that someone was most likely Marine Shale. Other reporters in town, state regulators and even other environmentalists I've spoken to all came to the same conclusion.

As of now, Rollins is shut down and so is Marine Shale. The public relations group is either defunct or has moved out of town, and Ossie Brown no longer represents Christians for Good Government, which itself seems to have vanished.

If they weren't a front for Marine Shale to directly oppose Rollins, then I guess the group must believe the government has become good enough, and they are no longer needed.

I recently met with Ossie Brown, who still practices in Baton Rouge, and had a nice conversation with him about Christians for Good Government. He told me in no uncertain terms that my conclusions were wrong. He told me that at a time when he wanted to enter the environmental law arena, an old campaign worker recruited him for the Christians for Good Government. Ossie said he was usually unpaid or paid only with what the group called "soup money" - a few dollars raised by pitching in at the group's soup dinner.

However, Ossie said he did not know about the public relations group or the young man who managed the protests. I personally find that a little odd. I've enjoyed a lot of pot-luck dinners in my day in a lot of churches and community centers. It seems to me that someone over soup dinner might have at least made a comment about the nice young man helping them protest.

The problem that we reporters had was that we were in the midst of a growing roster of environmental groups and had a difficult time distinguishing earnest, legitimate groups from Johnny come lately's, front groups or even misguided individuals seeking TV time. Often, existing environmental groups didn't actively deny or confirm the sincerity or constituency of the new groups or individuals. So actions by front groups, which seemed to get passive support of local environmentalists, hurt the credibility of local environmental groups when reporters painted them all with the same brush of skepticism. There were times when neither I nor my colleagues knew if we were covering legitimate environmental groups and concerns or not. That hurt credibility among reporters of all environmental groups. Although I can't say these fronts hurt membership in the environmental movement, I can say they caused reporters to think twice about covering yet another story created by yet another environmental group of questionable origin.

Lately some press has highlighted front groups. Associated Press writer Jim Drinkard produced a story late in 1997 which ran under headlines across the country similar to this: "Citizens Groups often Fronts for Powerful Financial Interests." Drinkard listed several organizations with grassroots sounding names like Northwesterners for More Fish, a group funded by Washington state industries. Too bad I or anyone else didn't reveal as much during the height of the environmental movement. If we had, we might have cleared the air on specific groups. As it turns out, the growing, general doubt about group origins took a lot of wind out of the movement's sails and brought into question the legitimacy of many aspects and people in the movement.

# Wrap Up

All good public speakers use this outline: 1. Tell the audience what you're going to say, 2. say it, 3. tell your audience what you just said. That's not exactly how you write a book, but I would still like to wrap things up a bit so you can draw your own conclusions. Here are mine.

The grassroots environmental movement I covered as a reporter is dead. Gone are the throngs of moderate citizens who attended enviro rallies and actively pressured polluters to clean up. Few politicians campaign on clean air, and few news stories cover the topic other than the obligatory time and space devoted to scattered protests against one company or another. I believe this is because the movement lost its direction and its credibility. National trade unions, civil rights groups, environmental extremists, conservation groups, even workers rights' groups fought to control the agenda, gain the membership and collect the dues. As in politics, all environmental issues are local, so when the local element was drowned out by national concerns of labor, civil rights groups, Greenpeace, and the Audubon Society, the grassroots movement got stranded. The current state of environmentalism is little more than a scattering of local skirmishes led mainly by national groups which swoop down to do battle against individual, local industrial sites, farming activities, logging companies or any industrial activity with someone to point at and blame.

Also, data slinging and extremist tactics, though good for a few headlines and a few days of attention, rang hollow after the fact and sent many moderate and conservative members of the movement looking for more realistic and worthwhile pursuits. While Greenpeace members continued to chain themselves to chemical plant gates, grassroots grannies and enviro moms were re-discovering soccer and baseball. Eventually, even Greenpeace had to close its offices.

I also believe at least one grassroots message ultimately hurt the movement. The difference between the messages "We can all help," and "We are all the problem" may be subtle but important. As soon as the grassroots movement convinced us we were all the problem, corporate America had no reason to continue to bear the lion's share of the responsibility for cleaning up. As a result, corporate America started calling for more citizens to do their part, which took some of the burden for pollution off corporations. (Simply note many of the ozone reduction laws were aimed at hairspray and car air conditioner, not the millions of tons of emissions coming from smokestacks.) The message might have better served the movement had it been simply "We can all help." With that message, leaders could still rally members to picket lines, corporate meetings and the halls of legislature to put pressure on polluters. At the same time, the movement could continue to instill a sense of personal responsibility for reducing, re-using and recycling without actually blaming the general public as the cause of pollution.

It seems to me the best way to revitalize the movement would be to return to local environmental issues linked nationally through living room meetings, idea swapping, networking and information trading of the early 80's movement. Local groups should look warily on national interests and national groups. Chase out the labor unions, civil rights groups and conservation groups, unless they buy into the local environmental agenda. Locals have bought into the union and civil rights agenda for too long. The current strength of the environmental agenda lies in pockets across the nation. The true national movement has broken down into isolated skirmishes against one company or another, but in those skirmishes lie the resources, people and ideas to renew the environmental movement on a larger level. The reason they are isolated skirmishes is for the reasons I've stated. The current issues appear to affect only small groups of people: minorities and the economically disadvantaged who live next door to the "polluter of the month." Large, effective grassroots movements need moderates, the middle class and soccer moms. If clean air and clean water were the true goal of current environmentalism, rather than vague notions of justice, then everyone would still have a stake in it. When the goal is merely choice jobs, benefits and community development for a few people living near a proposed plant, then the vast majority of the nation has no real stake and therefore no real interest - too many other clearer issues are pressing.

One day, leaders of the current, shrunken movement may stick their heads up and discover they are devoting lots of time and energy toward their cause, yet have drifted away and become isolated from middle America. Those leaders will need to break out the oars and start pulling back toward the mainstream if they hope to revitalize their movement.

Glen Duncan
Baton Rouge, LA

# Sources

The following sources are in addition to my own observations and notes gathered as a reporter covering the environmental beat

## A Simple Movement Is Born and Goes to Washington

Carson, Rachel; *Silent Spring*, Greenwich, Conn., Fawcett, 1962, 344 p.

Hays, Samuel P*.; Environmental Politics in the United States*, 1955-1985; Cambridge University Press, Cambridge, NY; 1987, 630 p.

Leopold, Aldo; *A Sand County Almanac*, NY, Oxford University Press, 1949, 226 p.

Shalley, Justin; personal conversation October 1, 1999

Switzer, Jaqueline Vaughn; *Environmental Politics - Domestic and Global Dimensions*, St. Martin's Press, NY; 1994, 379 pp.

## The Death of a Movement

Nelson, Sen. Gaylord; Speech given to Catalyst Conference in Illinois in 1980, as re-printed in CNN Interactive, Cable News network, April 21, 1996.

Orr, Marylee; Letter in LEAN NEWS, Regional Mini-brief, upper River Corridor Region, September 1995.

Ward, Bud; *Ink, Air Time for Environment: Endangered Species?* in Environment Writer, v.8 n.8, November 1996, p. 3.

_____ ; *Network News Environmental Coverage Down By Two-Thirds Since 1980*, in Environment Writer, v.9 n.6, September 1997, p. 1.

**The Price of Success - How the Movement Died**

O'Neill, Speaker Tip; *All Politics is Local and other rules of the game*, NY, Time Books/Random House, 1994, 190 p.

**Labor Goes Green**

Locked Out - The Story of OCAW Local 4-620, informational video, produced by Organizing Media Project - Oil, Chemical and Atomic Workers International Union, Washington, D.C., circa 1988

Schwab, Jim; *Deeper Shades of Green: The Rise of Blue-Collar and Minority Environmentalism in America*, Sierra Club Books, 1994, 480 pp.

## Don't Mention Birds!- Conservationists (and everyone else) Go Green

All Things Considered; news coverage of closure of Greenpeace offices, National Public Radio, Washington, D.C., Sept. 15, 1997

Arnold, Ron and Alan Gottlieb; Trashing the Economy - How runaway environmentalism is wrecking America, 2nd ed., Free Enterprise Press, Bellevue Washington, 670 pp.

Boerner, Christopher and Jennifer Chilton-Kallery, Nature groups must battle themselves, syndicated column, Knight Ridder Newspapers, Dec. 18, 1994 as it appeared in The Sunday Advocate, Baton Rouge, Louisiana, p.9B

Cockburn, Alexander, editorial column, The Nation, v. 265, n. 10, Oct. 6, 1997

Dowie, Mark; *Losing Ground: American environmentalism at the close of the twentieth century*, MIT Press, 1995, 317 pp.

Foreman, Dave; *Confessions of an Eco-Warrior*, Harmony Books (Crown), 1991, 228 p.

Sale, Kirkpatrick; The Green Revolution 1962-1992, Hill and Wang, 1993, 124 p.

Switzer, Jacqueline; *Green Backlash, The history and politics of environmental opposition in the United States*, Lynne Rienner Publishers, 1997, 323 p.

Werbach, Adam; *Act now, Apologize Later*, Cliff Street Books, 1997, 307 p.

Wiltshire, Susan D. and The League of Women Voters Education Fund, *The Nuclear Waste Primer - A Handbook for citizens*, revised edition., Lyons and Burford, New York, 1993, 170 pp.

## Green Justice- The Civil Rights Movement Crowds Out The Environmental Movement

Advocate, The; various articles, Capital City Press, Baton Rouge

Dunne, Mike; *Environmental justice questions unanswered*, The Advocate, Capital City Press, Baton Rouge, Tuesday, October 20, 1998

Clinton, President William J.; *Federal Actions to Address Environmental Justice in Minority Populations and Low-Income Populations*, Executive Order 129898, Feb. 11, 1994

Commission for Racial Justice; *Toxic wastes and race in the United States: a national report on the racial and socioeconomic characteristics of communities with hazardous waste sites*, United Church of Christ / NY,

Public Data Access: Inquiries to the Commission, 1987, 69 p.

Foreman, Christopher; *The Promise and Peril of Environmental Justice*, Washington, D.C., Brookings Institution, 1998, 191 p.

Ostheimer, John M. and Leonard G. Ritt; *Environment, energy and black Americans*, in Sage Research Papers in the Social Sciences (Human Ecology Subseries n. 90-025, v. 4) Sage publications, Beverly Hills, 1976

Soto, Tom; Clean air a right, not an amenity; the environmental movement won't succeed without the urban poor and people of color, editorial in Los Angeles Times, April 11, 1993

Swan, J.A; *Public Response to Air Pollution*, in Environment and the Social Sciences: Perspective and Applications, J.F. Wohwill and D.H. Carson, eds., American Psychological Association, Washington, p 66-74, 1972

Hampton, Henry and Steve Fayer, Eyes on the Prize, Blackside, Inc., Boston, MA, PBS Video, 1986

**Global Green - A New Social Order**

EnviroScan #136, http://www.hookup.net/rsirvine/, Public Relations Management Ltd., 7/21/96

Hentoff, Nat; Princeton professor advocates infanticide, editorial column, Newspaper Enterprise Association in The Advocate, Baton Rouge, LA Sept. 15, 1999

List, Peter C., *Radical environmentalism - Philosophy and Tactics*, Belmont, CA, Wadsworth Publishing, 1993, 276 p.

Quotes from Prominent Environmentalists; http//off-road.com/green/ecoquote.html, 7/21/96

Singer, Peter; *Practical Ethics*, NY, Cambridge University Press, 1979, 237 p.

_____; *Should the Baby Live - the problems of handicapped babies*, NY, Oxford University Press, 1985, 228 p.

Williams, Walter; editorial column, Heritage Features Syndicate, June 25, 1992

The Advocate, various articles, Capital City Press, Baton Rouge, LA.

## Other Things That Hurt the Movement

Drinkard, Jim; *Citizens Groups Often Fronts for Powerful Financial Interests*, Associated Press in The Advocate, Baton Rouge, p. 14A, December 18, 1997.

ECI; Soil sampling activities/results; report copy provided by Dow Chemical

Fialka, John. *Superfund Ensnares Thousands of Small Firms in a Legal Nightmare, Fueling Overhaul Drive.* The Wall Street Journal, p. A20, March 19, 1997

Rainbow Educational Video. *"Reducing, reusing, and recycling: Environmental Concerns"*

Schnaiberg, Allan and Kenneth Alan Gould; *Environment and Society - The Enduring Conflict*, St. Martin's Press, NY, 1994, 255 p.

Seattle, Chief 1790-1866; *Brother Eagle, Sister Sky: A Message from Chief Seattle*, paintings by Susan Jeffers, 1st ed., New York, Dial Books, 1991, unpaged.

Stott, Jon C.; *Native Americans in Children's Literature*, Oryx Press, Phoenix, AZ, 1995, 239 p.

# Index

## A

ABC, 2, 33
ACLU, 31, 44, 79
action teams, 58, 59, 69
Addis, 136
Advocate, The, 33, 178, 179, 180
AFL-CIO, 31, 44
agenda, 3, 22, 23, 29, 45, 50, 60, 61, 74, 80, 81, 83, 84, 93, 96, 98, 102, 108, 110, 111, 112, 116, 141, 159, 171, 173
Agents Provocateur, 159
Alsen, 14, 69, 167
Amelia, 60, 130, 160
Amery, Carl, 113
antifreeze, 18
Arnold, Ron, 76
Ascension Parish Residents Against Toxic Pollution, 15, 48
Associated Press, 169
Athens, 5, 120, 121, 122
Athens Convention and Visitors Bureau, 121
Audubon Society, 54, 55, 56, 72, 75, 77, 171
Audubon Zoo, 39
AWARE, 95

## B

bacteria, 29
Barri, Judi, 112
BASF, 46, 47, 48, 49, 50, 51, 52, 64, 65, 76, 132
Baton Rouge, 2, 5, 13, 14, 15, 26, 33, 40, 47, 48, 49, 50, 51, 60, 61, 63, 67, 69, 70, 71, 76, 88, 100, 102, 131, 137, 160, 161, 163, 166, 167, 168, 174, 177, 178, 179, 180
beat, environmental, 2, 26, 27, 29, 31, 175
Beluga, 63, 65, 66, 68, 76
Berle, Peter A.A., 56
Bhopal, 48, 50
black caucus, 91
Brenner, Yul, 117
Brower, Dave, 113
Brown, Ossie, 167, 168

Browning, Carol, 27
Bryant, Pat, 57, 96
Bush, George, 22

## C

Calcasieu, Lake, 32
cancer, 13, 16, 17, 93, 107, 112, 130, 133
Cancer Alley, 23, 24, 69, 70, 71, 132, 134
candidates: congressional, 19
carcinogens, 13
Carson, Rachel, 8
Center for Media and Public Affairs, 30
CERCLA, 150, 152, 154, 158
Chase, Steve, 112
Chavis, Ben, 79, 80, 107
Chen, Vivien, 130
chlorinated hydrocarbon, 139
chlorine, 29, 130
Christians for Good Government, 166, 168
Citizen Action, 42, 44
Citizens against Nuclear Trash, 15

Citizens for a Clean Environment, 14
Civil Rights Act, 99, 100
Clarke County, 120
Clean Air Act, 7, 9, 22
Clinton, 20, 22, 23, 83, 99, 101, 132, 134
Coalition for Clean Air, 80
communities, minority, 80, 82, 101, 102, 104, 106, 126
Comprehensive Environmental Response, Compensation and Liability Act, 150
congress, 151
conservation groups, 3, 44, 54, 55, 57, 141, 171, 173
conservationists, 6, 55
Convent, 87, 88, 90, 92
creep, extreme, 117, 118, 123, 124

## D

Data Slinging, 129
Department of Environmental Quality, 49, 85, 92, 99, 105, 139, 162

Deroussel, Martin, 65
Detjen, Jim, 25, 30
diapers, 144. *See*
Dickerson, Janice, 85, 108
disenfranchisement, 86
Donora, 7, 8
DOW, 71, 137
Dow Chemical, 18, 136, 137, 139, 140, 180
Dowie, Mark, 77, 177
Drinkard, Jim, 169, 180
Dudley, Barbara, 74
dumps, municipal, 153
Dunne, Mike, 102, 178

**E**

Earth Day, 6, 7, 8, 9, 22, 24, 32, 39, 40, 41, 80, 116, 142
Earth day, Anti-, 40
Earth Fest, 39, 40
Earth First!, 44, 74, 110, 112, 116, 120
Earth Night Action Group, 116
Eastman Kodak, 12
ecofeminism, 111
ecology, deep, 111
ecology, shallow, 111
ecotage, 111, 115

enviro moms, 108, 122, 172
enviro police, 111
environmental, 51
Environmental Protection Agency, 8, 83, 90, 98, 101, 150, 162
environmentalism, 6, 7, 43, 47, 54, 77, 84, 171, 173, 177, 179
environmentalists, 3, 5, 7, 8, 10, 16, 17, 19, 20, 23, 24, 28, 30, 43, 44, 50, 55, 72, 84, 86, 87, 89, 91, 96, 98, 110, 129, 134, 141, 144, 145, 153, 159, 160, 164, 168, 169
Environmentalists, 17, 125, 162, 180
enviros, moderate, 128
EPA, 13, 27, 84, 85, 86, 87, 88, 90, 91, 92, 94, 95, 98, 99, 100, 102, 104, 150, 151, 152, 153, 154, 156, 157
equity, environmental, 30, 81, 84, 85, 86, 87, 88, 91, 93, 94, 95, 98, 99, 100, 104, 105, 106, 107, 108, 124
Erlich, Paul, 112

ethylene glycol, 18
Executive Order 12898, 83
extremism, 29, 75, 76, 122, 124
extremists, 3, 24, 29, 30, 32, 37, 39, 40, 42, 74, 111, 113, 120, 122, 123, 127, 129, 130, 131, 133, 135, 136, 141, 171
Exxon, 146
Eyes on the Prize, 96, 97, 179

**F**

Fayer, Steve, 96, 179
Fialka, John, 155, 157, 180
Fletcher, Arthur A., 104, 105
Flint, 94
Fonda, Peter, 69
Foreman, Christopher, 100
*Foreman, Dave*, 74, 113, 115
Foster, Gov. Murphy "Mike", 88, 90, 121, 140
freon, 38, 143

Friends of the Environment, 15
Friloux, Jim, 99, 100, 105

**G**

Gaudet, Kay, 13, 134, 135
Geismar, 46, 50
Georgia Gulf, 63, 64, 65, 66, 67, 68
Georgia, University of, 120
Germany, 43, 46, 51, 113
Gettysburg, 155, 156, 157
Gibbs, Lois, 9, 10
Global Wildlife Fund, 54, 72
GM, 146
Golsby, Marsanne, 141
Gore, Al, 112, 133
Gottlieb, Alan, 76, 177
grannies, grassroots, 17, 23, 31, 32, 33, 46, 108, 145, 158, 172
Gray, Bob, 63, 66
Great Toxic March, 33, 51, 69
Green Party, 51, 110, 113

green, ultra, 114, 115, 121
Greenpeace, 26, 31, 40, 42, 44, 50, 51, 55, 58, 59, 60, 61, 62, 63, 64, 65, 66, 67, 68, 69, 70, 71, 72, 73, 74, 75, 76, 86, 87, 88, 89, 92, 95, 110, 115, 117, 118, 120, 126, 127, 128, 132, 140, 162, 165, 171, 172, 176
Guelph, 111, 112
Gulf Coast Tenants Association, 40, 96

## H

Hampton, Henry, 96
Hazardous Waste Treatment Council, 164
Hispanics, 78
Hooker Chemical, 9
Hopper, Dennis, 69
Hunt, Brian, 67

## J

Jackson, Jesse, 87, 93, 95
Jackson, William H., 120
Japan, 117

Jeffers, Susan, 147, 148, 149, 181
journalism, 2, 25, 30, 121, 137
Junior Ladies Garden Club, 121
justice, environmental, 80, 82, 83, 84, 85, 86, 87, 89, 90, 91, 92, 94, 98, 101, 102, 103, 104, 105, 108, 109, 124, 125, 126, 127
justice, green, 78, 178
justice, Presidential commission on Environmental, 44
justice, racial, 79, 80, 178
justice, U.S. Department of, 44, 86, 155

## K

Keller, Bill, 74
Kent, Jack, 165
Kirkland, Les Ann, 11
Kuehn, Bob, 124, 125, 126, 127

## L

labor, 34, 46, 47, 85, 132, 171, 173

leaders, black, 79, 80, 81, 87
League of Women Voters, 57, 178
LEAN, 14, 25, 49, 92, 176
Leopold, Aldo, 8, 175
Liebman, John, 64
Lincoln, Abraham, 11
List, Peter C., 111, 179
lock-out, 47, 48, 132
Los Angeles, 80, 179
Louisiana Chemical Association, 62
Louisiana Environmental Action Network, 14, 25, 49, 85, 87
Louisiana State University, 107, 130, 133
Love Canal, 5, 9, 10
Love, William T., 9
Loyola, 93

**M**

Malek-Wiley, Darryl, 57
manifest destiny, 7
Marine Shale, 13, 32, 59, 160, 161, 162, 163, 164, 165, 166, 168

Marine Shale Processors, 13, 160
media, 8, 12, 14, 24, 30, 36, 41, 42, 43, 44, 46, 47, 49, 50, 52, 61, 62, 67, 88, 100, 123, 129, 132, 133, 140, 159, 163, 165, 166, 167
methyl ethyl ketone (MEK), 137
methylene chloride, 138
minorities, 78, 81, 82, 84, 95, 97, 98, 101, 106, 124, 125, 173
miscarriages, 13, 130, 134
Mississippi River, 5, 13, 26, 47, 51, 59, 60, 63, 64, 65, 68, 70, 101, 128, 130, 131, 144
Monongahela, 8
Morgan City, 13
movement, civil rights, 41, 79, 84, 90, 96, 105, 108
movement, environmental, 2, 3, 5, 8, 10, 14, 20, 21, 22, 24, 25, 26, 28, 29, 31, 35, 37, 38, 39, 41, 44, 45, 46, 48, 50, 51, 52, 54, 55, 57, 58, 59,

61, 73, 76, 77, 78, 83, 88, 89, 90, 95, 96, 97, 98, 103, 106, 108, 109, 110, 115, 122, 123, 141, 157, 160, 169, 170, 171, 173, 179
Murray, Bill, 136

**N**

NAACP, 31, 42, 44, 79, 80, 87, 89, 90, 91, 92, 93, 107
National Centers for Disease Control, 133
National Environmental Justice Advisory Council, 100
National Public Radio, 74, 113, 176
Nature Conservancy, 54, 72, 77
Nelson, Senator Gaylord, 22
neuroblastoma, 13, 134
New Orleans, 2, 5, 13, 33, 39, 40, 50, 51, 57, 69, 96, 102, 104, 131
news media, 31, 165
Newspapers, 19, 48, 75, 177
Niagara Falls, 5, 9

Nicholson, Jack, 69
Northwesterners for More Fish, 169
nuclear tests, 58

**O**

OCAW, 47, 48, 49, 50, 51, 60, 176
Oil, Chemical and Atomic Workers union, 46
O'Neill, Congressman Tip, 35, 176
Orr, Marylee, 14, 15, 84, 92, 176
Ostheimer, 78, 179
ozone, 15, 38, 39, 142, 143, 172

**P**

Pacific Gas and Electric Company, 116
Peabody Award, 121
Peabody, George Foster, 121
People for the Ethical Treatment of Animals, PETA, 11, 110
Perot, Ross, 36
Philadelphia Inquirer, 25
Pittsburgh, 7

Plaquemine, 11, 18, 23, 63, 94, 126, 136, 137, 138
politician, 20
Pontchartrain, Lake, 32
Potentially Responsible Parties, 151
practices, best available, 17
President, 10, 22, 23, 36, 83, 84, 86, 99, 101, 105, 112, 132, 134, 178
press, 21, 23, 50, 52, 57, 60, 82, 89, 90, 95, 97, 103, 126, 128, 132, 140, 160, 164, 166, 169
Price, Miriam, 13
PRPs, 151, 152, 157
pseudoenvironmental, 72
PVC, 140

## R

racism, environmentalracism, 79, 80, 81, 87, 91, 92, 98, 99, 107, 108, 124
*Rainbow Warrior*, 58
recycle, 18, 19, 26, 32, 38, 142
recycler, hazardous waste sham, 162, 163
Resource Conservation and Recovery Act, 18
rights, civil, 3, 30, 31, 34, 40, 41, 44, 57, 79, 80, 81, 82, 83, 84, 86, 87, 90, 93, 94, 95, 96, 97, 98, 102, 103, 104, 105, 106, 107, 108, 109, 141, 171, 173
risk, 16, 17, 47, 102, 116, 117, 163
Ritt, Leonard G., 78, 179
Rochester, 5, 12
Rollins Environmental Services, 14, 160
Romeville, 92, 126

## S

Sale, Kirkpatrick, 77
Santa Barbara, 8
Seattle, Chief, 147, 148, 181
SEJ, 25, 27
Select Steel Corporation of America, 94
Shalley, Justin, 8, 175
Shell Oil, 40

Shintech, Inc., 87, 88, 89, 91, 92, 93, 94, 95, 101, 125, 126, 127
Sierra Club, 49, 54, 55, 57, 72, 75, 76, 176
*Silent Spring*, 8, 175
slinging, data, 24, 129, 131, 132, 136, 140, 149, 171
social order, new, 31, 110, 120, 125
Society of Environmental Journalists, 25, 30
Soto, Tom, 80, 179
South Louisianians Against Pollution, 14
Spain, John, 26, 67
St. Gabriel, 13, 130, 134, 135
St. James Citizens for Jobs and the Environment, 95
St. James Parish, 86, 87, 90, 91, 93, 105
Stevens, Darrell, 50
Stevens, Gov. Isaac, 147
Stott, Jon C., 148, 149, 181
SunnyRay Cafe, 157
superfund, 150, 153, 158

Superfund, 3, 10, 50, 144, 149, 151, 152, 154, 155, 156, 157, 158, 180
Supreme Court, Louisian, 127
Swan, J.A., 78, 179
Switzer, Jacqueline, 77, 175, 177

**T**

Teach-Ins, 22
technology, best available, 17
technology, maximum achievable control, 17
tetrachloroethene, 138, 139
Title VI, 92, 99
totalitarianism, 110, 111
Toxic Tours, 50
Toxic Watch Program, 50
trade unions, 3, 30, 42, 44, 54, 106, 141, 171
Tree that owns itself, The, 123
Tulane University, 13, 87, 89, 103, 124, 130, 135
TV, television, 1, 2, 6, 12, 19, 23, 24, 26, 33,

36, 37, 41, 59, 60, 61, 62, 66, 68, 71, 119, 128, 132, 133, 135, 137, 160, 169

## U

U.S. Commission of Civil Rights, 104
U.S. Steel, 7
Uncle Sam, 153
United Church of Christ, 79, 178
United States vs Keystone, 155, 156, 157
USSR, 117

## V

Vietnam War, 8

## W

WAFB, 33
Wall Street Journal, 155, 180
Washington, D.C., 23, 176, 177, 178
waste reduction, 19
waste stream, 17, 18
water, chlorinated drinking, 29
Watergate, 8
Werbach, Adam, 76, 177
whaling, 59, 72, 117
White House, 16, 23
White, Dr. LuAnn, 103, 130, 135
Williams, Barbara, 155, 156
Williams, Walter, 112
Winfrey, Oprah, 23
Working Group for Environmental Justice, 83

## Y

Yamada, Gerlad H., 100

## About the Cover

After searching through stock photos for days, I grabbed by camera and set out on Dawson's Creek in Baton Rouge. An old, wise hiker once told me look back over your shoulder now and then. I did, and I'm glad, because there was my cover. There is no stronger symbol of middle America than the shopping cart, and no clearer indication of a waning environmental movement than junk in an urban creek within reach thousands of everyday citizens who could stoop to clean it up.